Ryan BROUGHAMS and Their Builders

by WILLIAM WAGNER

PUBLISHED BY

HISTORICAL AVIATION ALBUM

P.O. BOX 33
TEMPLE CITY, CALIF. 91780

ACKNOWLEDGMENT

No author can ever adequately thank all who have contributed significantly to an historical work such as "Ryan Broughams and Their Builders."

An alphabetical listing of names somehow always leaves out a number of individuals who have helped. Nor does such a list give full credit to those whose special interest and effort has been particularly useful.

All who have helped are aware of their contributions and we extend them our grateful appreciation. However, we do feel compelled to single out special credits.

The late H. A. (Jimmy) Erickson was responsible for most of the fine documentary photo coverage of the Ryan Airlines organization 1925-1929, including the Lindbergh period.

Pioneer Ryan employees have contributed generously from their personal files. Among these are John van der Linde, Dan Burnett, Bill Monday, Ed Morrow and Frank Wiley. Members of the Bowlus, Hall, Edwards, Ayars, Tindale and Mirow families have made available new historical material.

Among those still living who were part of the Brougham story and have contributed much to this volume are Art Mankey, Bob Adams and Tom Mathews. The latter in particular has been most helpful as he also was in preparation of a prior work, "Ryan, the Aviator."

Bud Gurney, airmail pilot with Slim Lindbergh on the St. Louis-Chicago run, has provided background information and photos, including those on which the cover painting by Matt Giacalone is based.

Much of the information regarding Brougham production in St. Louis was unobtainable until, by good fortune, we came in touch with Hugh Belknapp, an employee who had thoughtfully retained many documents and photographs which enabled us to complete this phase of Brougham history. Also helpful in compiling the St. Louis story were Gerry Balzer and George Brooks of the Missouri Historical Society. Ev Cassagneres, Ryan historian of the American Aviation Historical Society and founder of the National Ryan Club, was always most cooperative, as were John Underwood, Vic Stuhr, Warren Shipp, Dave Hatfield and Bob Peck to name only a few of the aviation buffs who shared research material.

To O. R. McNeel we are especially indebted for the story of the Ryan Mechanics Monoplane Company and to Martin Jensen for the lively, often humorous story of Leo the Lion.

And, lastly, it should be noted that without the interest Paul Matt has long shown in Historical Aviation, his Album would not have been published.

WILLIAM WAGNER

San Diego, California
May 23, 1974

LIBRARY OF CONGRESS
CATALOG CARD NUMBER
74-79324

ISBN No. 911852-76-X

Typography
TRADE TYPE
San Gabriel, California 91776

CHAPTERS

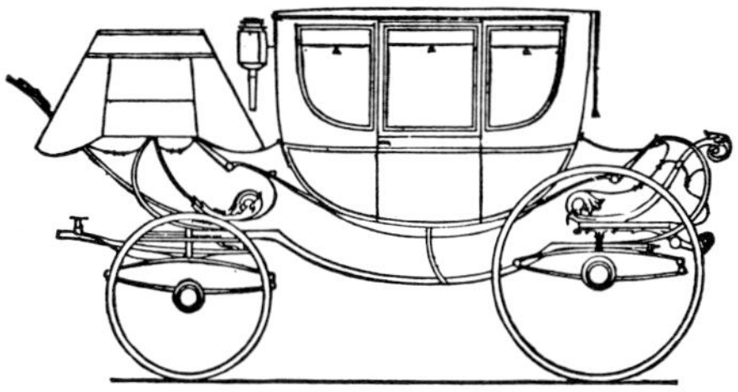

Foreword

LOLA MONTEZ, the classy 19th century courtesan, was the kind of spirited woman who would have appreciated the speed and comfort of the Ryan Brougham cabin plane had it been available in the days she ricocheted between the social and power capitals of Europe – not to mention California.

Named after one of her many benefactors, Lord Henry Peter Brougham, British barrister and philosopher, the designation "brougham" for a vehicle was first used in 1838. It described the "closed town carriage with seats inside for two or four persons," which Lord Brougham designed and had built for his own use.

With the turn of the century, the horseless carriage made its debut and sorely needed descriptive names for the various automobile body styles were naturally based on horse-drawn carriages of the day. There were the phaeton, cabriolet, landau, victoria and coupe to cite but a few examples.

It was Ransom E. Olds, of Oldsmobile and Reo fame, who was the first to apply the name "brougham" to an automobile. An early Oldsmobile advertisement which called attention to the company's 15 years' experience, listed the product lines for the new model year.

The newly named Brougham, powered by a 9-horsepower vapor (gasoline) engine, was featured in both two- and four-passenger versions along with the four-passenger Cabriolet (vapor) and the Phaeton with either vapor or electric power.

Sixty-five years lapsed between Lord Brougham's "brougham" and that of pioneer automaker Olds. Another quarter century was to pass before Ryan Airlines, Inc., came along to perpetuate the "brougham" name in the aerial age.

It was November 1926 and Ryan Airlines was running proving flights on its first cabin plane, which employees had already nicknamed the *Bluebird* because of its light blue color styling.

Nine months earlier, February 14, T. Claude Ryan had made the first flight in his new M-1 open cockpit monoplane. He had designed it for use on the early contract air mail routes, but an occasional private owner also bought the snappy new plane. By Fall it was obvious that the M-1 would find a more receptive market if potential customers could also be offered a "closed carriage with seats inside."

H. A. Erickson

. . . the most famous airplane in history

The Ryan Airlines crew turned to the task and by November the *Bluebird* 5-place cabin plane was in the air. But would customers buy a *bluebird?*

Ryan's $15 - a - week public relations man, Tom Mathews, said he "knew the name would endear itself to love-sick youth and old ladies, but it wouldn't sell any airplanes. I told Claude Ryan and his partner, B. F. Mahoney — and sales manager J. B. Alexander — the reasons why an airplane-automobile tie-in might be more acceptable to customers than simply a new airplane name."

Thus the name Brougham was selected to appear in the first advertisement — January 1927 — for the new line of Ryan closed cabin planes. Four months later the Ryan Brougham was known around the world. Lindbergh's *Spirit of St. Louis,* a Brougham-type Ryan monoplane had become the most famous airplane in history.

During the five years the Broughams were built — and a decade after that — they were among the most prestigious planes in the sky.

This is their story.

WILLIAM WAGNER

San Diego, California
March 24, 1974

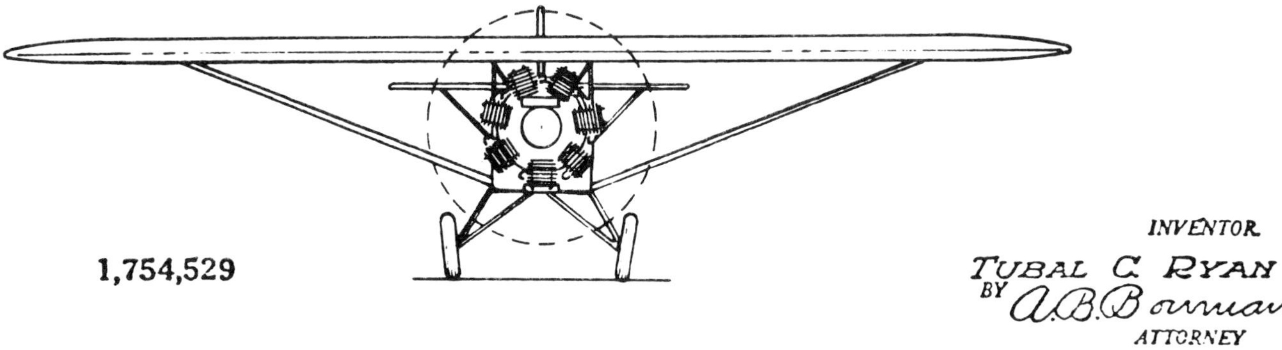

Making of a Monoplane

THE Brougham saga begins with T. Claude Ryan and the M-1 monoplane. From his training as an Army Air Service pilot; from studies as an engineering student at Oregon State College, and from practical experience as the owner-operator of Ryan Airlines, Inc., in the early 1920s, Claude felt that the monoplane was the design of the future.

He often discussed his concepts with Hawley Bowlus, his versatile shop superintendent, who had brought Claude's ideas for modification of surplus military planes to reality in the Ryan Standards used on the Los Angeles San Diego Air Line. Bowlus also helped Ryan rebuild Donald Douglas' first airplane, the *Cloudster,* from an open-cockpit sightseeing biplane to a 12-passenger enclosed cabin airliner for Ryan Airlines' Los Angeles run.

It was not until federal legislation was enacted in February 1925 enabling the Post Office to contract with private operators to carry U.S. Air Mail that Claude's ideas for a monoplane became economically feasible. By then he shared ownership of Ryan Airlines, Inc., with his partner, B. Franklin Mahoney, who had encouraged Claude to start the daily passenger airline service to Los Angeles.

The three of them – Ryan, Bowlus and Mahoney – decided to give serious consideration to the idea of developing a new monoplane especially for the privately operated airmail lines.

Traditionally new airplanes have their birthpangs on the back of an envelope. Claude made no such hackneyed claim for his new airmail monoplane but Tom Mathews insists that Ryan did indeed make the first sketches of his concept in the traditional manner.

"One morning in 1925," Mathews recalled, "Claude came in with a sketch he had drawn on the back of an envelope and showed it to Mahoney, who was enthusiastic. I know it was an envelope because I saw the sketch on Mahoney's desk when I took a letter in for him to sign. Next day Claude took the sketch and flew to Los Angeles to discuss it with an aircraft engineer."

Ryan knew the market was ripe for a new concept suitable for the expanding airmail service. He was convinced he could build a monoplane which would carry half again the pay load of the deHavilland biplanes used on the government-operated transcontinental airmail routes. And, he felt sure it could be done with only half the power of the big Liberty 400 h.p. engine in the wartime DH, and at 20 miles an hour higher cruising speed.

Ryan Aeronautical Library

T. Claude Ryan, Army Air Service pilot

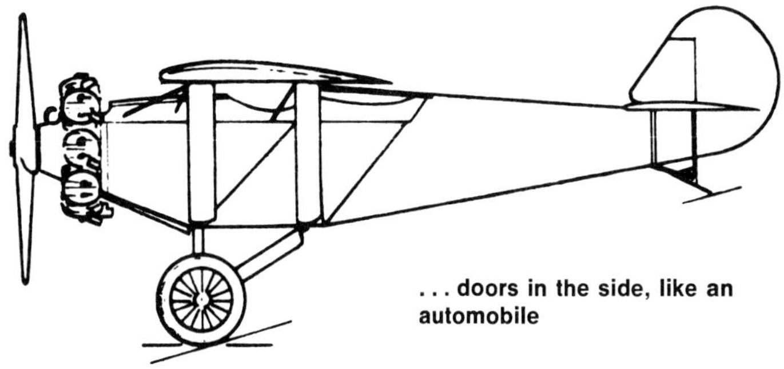

. . . doors in the side, like an automobile

Claude's idea was to build an externally braced monoplane of unusually sturdy construction and very high performance per horsepower. This would require a strong fuselage frame, with the single wing mounted atop this, leaving an area open between wing and fuselage to provide unobstructed visibility from the open cockpits, much as in a biplane. The new plane would have open doors in the side, like an automobile, for the pilot and two passengers. With two passengers and pilot, 300 pounds of baggage and mail could be carried. Strictly for mail use, payload would be 600 pounds.

Later Ryan patented his 'parasol' monoplane design and assigned rights to Ryan Airlines, Inc., without receiving any payment.

Looking back on the events of those days, Claude recognizes he was influenced by the design of many other planes of that period, but none more so than that of the Wright-Bellanca WB-1 strut-braced cabin monoplane. It had been designed in 1925 by Guiseppe M. Bellanca as a flying demonstrator to show the efficiency of Wright's new radial air-cooled J-4 Whirlwind engine.

"Pilots were not exactly confident of monoplane design in those days," Claude Ryan related. "One reason was that there had been a void of many years in new design because surplus World War I biplanes were available so cheap.

"We knew we'd have a great psychological handicap with a monoplane even though there was general agreement that the monoplane would ultimately rule the skies. But, unfortunately, there had been too many instances of experimental monoplanes having their wings fold in flight.

"Although I'd had engineering training, had drawn up the original configuration and figured out the basic design, I didn't consider myself a full-fledged aeronautical engineer. Since we didn't have a full-time engineer we decided to contract for the stress analysis and detailed drawings."

* * *

"Aircraft engineers you could count on your fingers," Mathews continued as he explained more about the Ryan monoplane.

"The Loughead (Lockheed) brothers, Douglas, Northrop, Waterman, and Waterhouse and Royer of Glendale. The latter seemed the best bet among California engineers so Claude took his sketch to them and outlined what he had in mind. They thought they could draw something up for a few hundred dollars down and sort of 'pay as you go' from there. Claude talked it over with Mahoney by phone and it was agreed so Ryan left a check and came back to San Diego.

"It was to be a strut-braced, open cockpit, high wing monoplane. The 36-foot wing was to be a full-span box-spar, mahogany plywood; Clark 'Y' airfoil, wood ribs, internally wire-braced with a spruce beam rear spar to attach the rear strut and aileron. Fuselage to be Shelby chrome molybdenum steel tubing, Z braced box type; split type landing gear, and tail skid. It was designed to use engines of up to 200 horsepower and with a 90 horsepower OX-5 it was supposed to cruise about 105 or 110 m.p.h. and land about 40 m.p.h. Claude specified safety factors of 8-to-1 fully loaded, equal to those for military pursuit planes.

"In a couple of weeks, after Hawley Bowlus had spent several days in Glen-

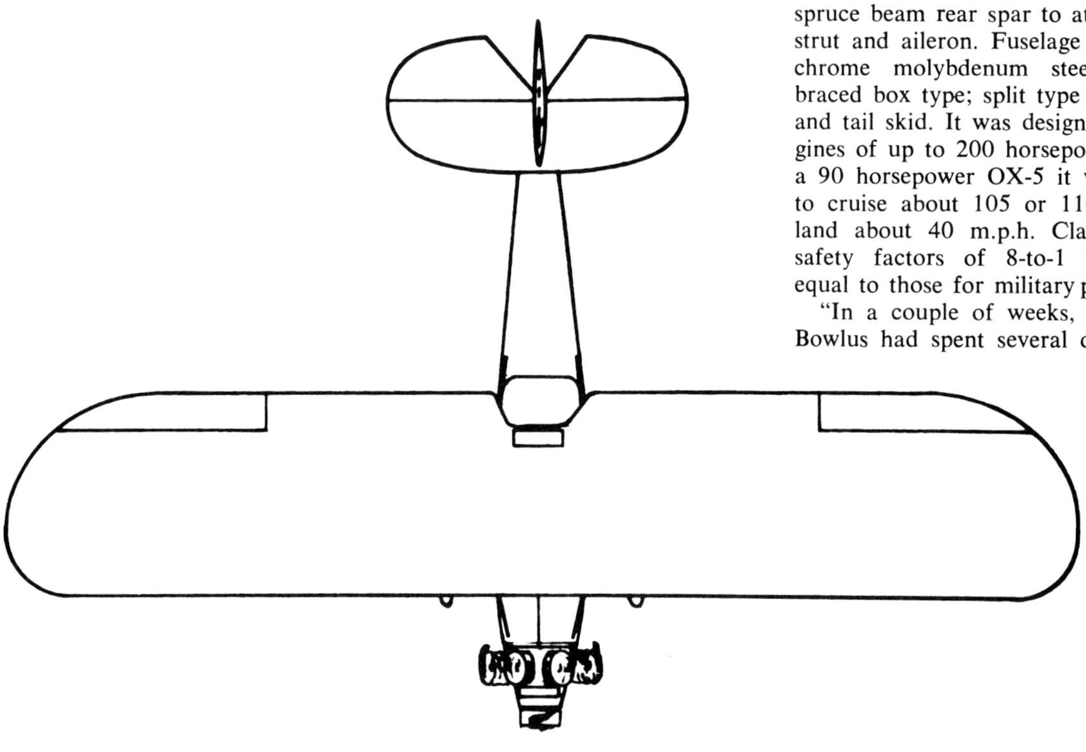

dale reviewing plans, a nice three-view drawing with general specifications arrived and Bowlus got together a small crew and on October 19, 1925 started building the plane which we decided to call the Ryan M-1. (M for monoplane, 1 for first model.)

"The wing drawings arrived fairly complete, also the fuselage drawing with tube thicknesses and type of weld specified. Everything went along in great style as we had plenty of experience with wood and fabric and some welding of tubing which a chap by the name of Mel Anderson was doing.

"About this time they shifted me up to Los Angeles to see if I could generate a little more business for the Los Angeles-San Diego Air Line. Meanwhile, J. B. Alexander, who had been hired as sales manager for Ryan Airlines in Los Angeles, started setting up a sales program for the Ryan M-1 (if and when).

"One morning George Allen, who piloted the airline schedule regularly, said Claude wanted me to go over to Waterhouse and Royer in Glendale and pick up some drawings as the men in San Diego had run out of work. I was to bring them down on the afternoon flight. When I got there I was a little surprised to notice a plane that looked just like ours in the process of construction, but not quite as far along. The wing wasn't complete, but the fuselage was covered and I think they had installed the landing gear.

"I introduced myself to William J. Waterhouse who introduced me to Lloyd Royer. My impression of Waterhouse was that he was a quiet type, more like an engineering professor; he looked like he knew what he was doing. I said, 'I came up for some more drawing. We're running out of work.'

" 'Did you bring a check from Mr. Ryan?'

" 'No, nobody said anything about a check, but I am supposed to take the drawings down on the afternoon flight and I will tell Mr. Ryan you would like to have a check.'

" 'No check, no drawings,' said Waterhouse and he went back to looking at some plans and talking to one of the workmen on the airplane. I had the feeling that he didn't have any more drawings, beyond those we already had and that he was building and drawing as he went along. I thanked them and said I thought Mr. Ryan would be disappointed and caught the afternoon flight to San Diego.

"When I told Claude about my failure to bring back more drawings and the reason, all he said was 'did you tell them we would send up a check as soon as we got the drawings?" But when I mentioned that Waterhouse and Royer had a duplicate airplane about as far along as ours, Claude talked to Mahoney and then went up to Los Angeles himself. I went out to the shop and talked to Bowlus about everything I had seen up there.

"In a few days Claude came back and went into a huddle with Mahoney and Bowlus and when Bowlus came out of the office he said, 'Okay, if you think I can do it.'

"So Bowlus and crew finished the airplane, using tried and true methods, improvising somewhat, following the three-view drawings pretty carefully, giving the tube thickness the benefit of the doubt by increasing size on the struts and landing gear to be sure we were safe. They welded up a motor mount to fit the 150 h.p. water-cooled Hisso A for the first test flight and when the big day arrived on February 14, 1926 Claude flew it and it was a big success. We flew about six weeks ahead of the Waterhouse and Royer monoplane.

Ryan Aeronautical Library

Hawley Bowlus, versatile shop superintendent

H. J. van der Linde

. . . the big day arrived on February 14, 1926 and Claude flew it . . .

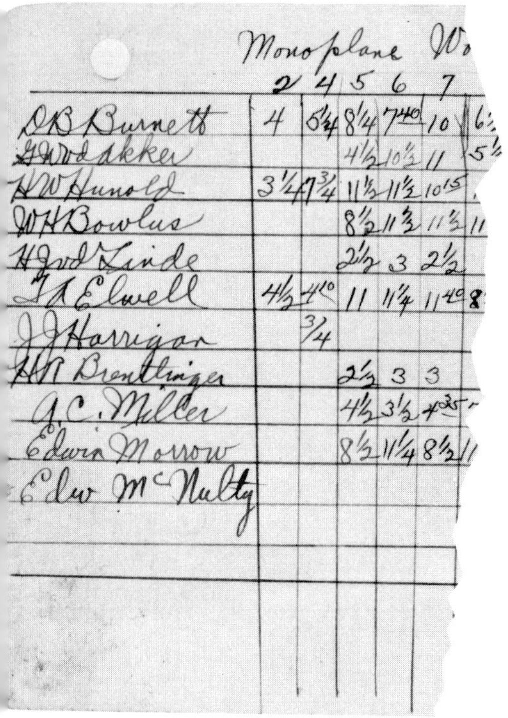

Monoplane W…	2	4	5	6	7	
D B Burnett	4	5¼	8¼	7 40	10	6
G.W. v d Akker			4½	10½	11	5½
H W Hunold	3½	7¾	11½	11½	10 15	
W H Bowlus			8½	11½	11½	11
H J v d Linde			2½	3	2½	
T A Elwell	4½	4 10	11	11¼	11 40	8
J J Harrigan		¾				
H R Brentlinger			2½	3	3	
J. C. Miller			4½	3½	4 35	
Edwin Morrow			8½	11¼	8½	11
Edw McNulty						

"Nobody ever told me what Claude said to Waterhouse or vice-versa but I always had the feeling that Claude reminded him of their agreement to pay as they went along. Claude was very easy to get along with, never showed annoyance or lost his temper but if anything could ever get to him I imagine it would have to be a challenge of his honesty.

"Certainly it was not a question of money, although Claude was not one to throw it around. We usually took in between $1500 and $3500 every week during the summer and early fall on our famous sightseeing flights over San Diego. Of course there were added expenses building the M-1 and then there was the suspicion that Waterhouse was going to finish his plane first and possibly sell it and the design to a competitor before giving us the final plans to finish ours.

"At any rate, Claude came back and turned the job over to Bowlus and told him to finish it up without further delay.

"So Bowlus, assisted by his father, C. D. Bowlus, and by John van der Linde, Ed Morrow, Dan Burnett and a handful of others finished the ship. It was surprisingly free of bugs."

As to Waterhouse copying the M-1 design, he was perhaps, in all fairness, doing only what is fairly standard practice among naval architects. When a boat-owner takes his ideas to a naval architect to have a custom design built, the architect retains the privilege of building the same basic design for other customers.

* * *

While overall performance of the M-1 was good, Claude recognized that the wing, though strong, was too heavy and expensive to build, and should be redesigned. Perhaps Donald Douglas, an old friend, would be willing to help by letting some of his engineers recheck and update the work of Waterhouse and Bowlus.

D. D. Hatfield

John K. Northrop . . . cleverest man on the staff

One day Claude flew up to visit Douglas at his tiny office on Wilshire Boulevard in Santa Monica. Could he suggest someone?

Yes, the chief draftsman, Art Mankey, would have some ideas.

"I had one fellow I rated the best of all," recalled Mankey. "His name was John K. Northrop and he was the cleverest man I had on the staff. Soon after that Jack and I began going to San Diego every weekend for several months to help Claude on his new monoplane."

The changes in the M-2 version were principally to replace the box-spar wing construction with an I-beam spar, design new ribs and incorporate other Douglas-type features. The change in the model nomenclature from M-1 to M-2 was as much a sales gimmick as for wing structure identification.

The water radiator for the Hisso-powered version had at first been placed in the leading edge of the wing but since that interferred with good airflow it was soon moved under the fuselage in front of the landing gear struts.

"Without the engineering help from Jack and Art," Claude often mentioned, "the M-1 might have been a clunker with all that extra weight. And the detailed stress analysis done by Northrop helped us out greatly later on when Professional Pilots Association, stimulated perhaps by Bill Waterhouse, questioned the safety of our design."

For airmail service the M-1 had been designed to use the 200 h.p. air-cooled Wright Whirlwind J-4B engine, but a quick-change engine mount made it possible to offer alternate power plants. For the initial test flight Claude had installed the French 150 h.p. Hispano-Suiza V-8 water-cooled engine, which, he explained, "like the OX-5, was available in abundance from war surplus stock at North Island, San Diego.

Ed Morrow

. . . with 150 h.p. OX-5

"The Hisso M-1 had such spirited performance compared with other aircraft that we decided to standardize on it and on the Wright Whirlwind 9-cylinder radial engine. We also made a couple of installations with the 120 h.p. Super-Rhone and while this model flew well, the engine was not too reliable."

One distinctive styling feature added by the shop people was the 'jeweling' effect achieved by burnishing the aluminum of the engine cowl, the wheel disc covers and the panel on top of the fuselage around the cockpits. For years this burnishing was virtually a Ryan trademark. Some said it had been ingeniously devised to hide the dents in the hand-hammered engine cowling!

On occasion an M-1 made the Los Angeles airline run or carried local sightseeing passengers but not every landing with passengers was letter perfect. Far from it, as explained with embellishment by Tom Mathews:

"About the shortest landing I ever saw was one by J. J. (Red) Harrigan, our chief pilot. With two passengers aboard he set it down so hard out of a nose high sideslip that the left axle snapped off and wheel and axle were sent fifty feet in the air as the M-1 dug in and spun around without nosing over.

"I gave the passengers a 'Certificate of Crash Landing' signed by Harrigan and charged them an extra $5 each to help pay for the damage. Claude didn't think I was being fair to them, but as I pointed out they paid for a crash landing — after the fact — and got it!"

* * *

Loening's case emphasizes the importance of timing as the deciding factor between success and failure.

In March 1918 near the end of World War I, Loening was asked to build a two-seat fighter which could exceed the British Bristol in performance.

To meet the requirement Loening designed the M-8 monoplane powered with a 300 h.p. Wright-Hispano engine.

"We decided," he said, "to go to the monoplane design, although we knew full well that it was in great disfavor and feared by pilots because of their suspicion of its weakness.

"I think this design was nearer to being inspired than anything I ever did."

The Loening M-8 turned out to have the sought-after performance exceeding that of the Bristol fighter. The second M-8 built went to McCook Field at Dayton (where Waterhouse was an engineer) for acceptance tests. The people there told Loening, "We are going to build 5000 of your planes on our next year's program."

H. J. van der Linde

. . . with 120 h.p. Super-Rhone

Agnes Allen Robinson

. . . with 150 h.p. Hispano-Suiza

H. A. Erickson

. . . with 200 h.p. Wright J-4B

In old-time aviation circles the discussion continues to persist as to whether Claude Ryan or Bill Waterhouse 'designed' the M-1. Some even contend that both Ryan and Waterhouse copied from Grover C. Loening. But Ryan says Bellanca's design ideas were a more important influence than those of Loening or Waterhouse.

One has only to look, for example, at the Douglas DC-8, Boeing 707 and Convair 880 to recognize that such similarities result from state-of-the-art design potential at any given time rather than from one copying the other.

There is credit for all for the monoplanes of the 1918-1927 decade. Ryan for his basic design concepts and ability to apply sound business principles in a young and uncharted industry. Waterhouse for his capability as a pioneer aeronautical engineer. Guiseppe M. Bellance for the efficiency of his Wright-Bellanca. Loening for his far-sighted vision and contributions to aviation spanning seven decades.

Then, very suddenly, the Armistice was signed and there were no production orders.

Everyone agreed that the monoplane was a design whose time would come, but it was more than seven years before Claude Ryan was to get his M-1 into production — at the right time.

There was indeed a similarity between the Loening M-8, the Ryan M-1 and the Waterhouse Cruzair. Clearly the similarity was state-of-the-art; not necessarily copy, although pioneer pilot Vance Breese recalled that Waterhouse had one set of drawings of the M-8 pursuit monoplane when he left Loening.

In any case, the United States Patent Office issued Patent No. 1,754,529 to T. Claude Ryan for ". . . aeroplanes, and more particularly the monoplane type." The drawing which accompanied the patent award was of his M-1.

RYAN AIRLINES, INC

MANUFACTURERS OF AIRPLANES

T. CLAUDE RYAN, PRES. | SAN DIEGO, CALIFORNIA

Tom Mathews

Partners B. F. Mahoney, left and T. Claude Ryan, right, give an M-1 sales pitch to prospect Vern C. Gorst, center

The First Order

THE longest and most difficult of the new airmail routes opened to private operation was that between Los Angeles and Seattle. Whatever plane was selected for that route was certain to attract attention and it would be a choice piece of business for the manufacturer.

One of the prospective operators of the Pacific Coast mail run was an Oregon bus operator, Vern C. Gorst, who wanted to respond to the Postmaster General's request of July 15, 1925 for bids on it and seven other routes. Late that year he came to San Diego to observe one of the few commercial airlines in being – the Los Angeles-San Diego Air Line. Ryan Airlines, Inc. had been operating the daily schedule since March first – a month after President Calvin Coolidge had signed the airmail legislation.

Gorst said he and those associated with him were interested in looking into the possibility of bidding on a new airmail route. Claude sized Gorst up as a real dreamer since he didn't look like a substantial businessman, but was not one to pass up any opportunity. Claude felt that he might get something out of Gorst's interest by teaching him to fly, pointing out that the best possible background a potential airline executive could have would be to know something about flying from an operational point of view.

Claude said he would be willing to advise Gorst and give him lessons at 'our regular rates', whereupon Gorst's response was, "Say, I might like to try that. How soon can I start?"

"In about five minutes," Claude replied.

As a result Ryan soon had Gorst in the air and subsequently gave him several hours of dual instruction over the next week or so, though Claude did not solo Gorst at the time.

Gorst's visit to Ryan's airport at Dutch Flats in San Diego coincided with the early stages of construction of the first Ryan M-1. Claude saw to it that Gorst had an opportunity to see the plane under development. He was able to point out its advantages and the improvements in the state of the art, and that the M-1 was ideal for Gorst's proposed contract airmail operations.

In about ten days Gorst returned to Oregon to see about forming a company to bid on the Pacific Coast airmail route, indicating to Ryan that he now considered Claude his technical adviser. Subsequently, Claude made the first flights in the M-1 and during the early testing phases Gorst stopped in several times to see the progress on the new monoplane.

Ryan not only gave Gorst flight instruction, he also helped him lay out the Los Angeles-Seattle route and prepare his proposal. There would have to be considerable night flying with departure northbound from Los Angeles at midnight and southbound from Seattle at 3:45 a.m. The tough Portland-San Francisco portion would be flown in daylight hours but beacons would have to be set up for the night flying portions between Los Angeles and San Francisco and between Portland and Seattle.

Three bids — two of them conditional — were received by the Post Office department, and Gorst was awarded the contract on January 1, 1926. A week later Pacific Air Transport was incorporated in Oregon by Gorst and business associates to operate the route.

As the M-1 was about to make its first flight, Gorst told Ryan that "your plane certainly looks the best — we will need six of them. But first I feel I have to make a tour of the United States to look at other planes which might be available and I want you to go along as my adviser."

Claude didn't feel he could leave the business at that critical time but did arrange for his sales manager, J. B. Alexander, to go along in his place. This took some convincing as far as Gorst was concerned, but he finally agreed that it was better to have Alexander as his adviser than to undertake the tour on his own.

Ryan explained to "J.B." that Gorst liked airplanes so much he would want to buy virtually every plane he would see during the tour. They were particularly concerned about the Curtiss Lark biplane which Gorst also favored. Consequently, Claude instructed Alexander that it was his assignment to bring Gorst back to San Diego without his prospect having contracted with anyone else for planes.

Admittedly it was an unusual arrangement for the potential customer — Gorst — to pay the expenses of an adviser whose sole assignment was to keep him from buying some competitors' equipment. (Gorst must have been serious about the Lark because the Aircraft Year Book for 1926 lists P.A.T.'s equipment as six Curtiss biplanes.)

The M-1 held its own, however, on merit. According to Alexander, "Casey Jones, handling sales for Curtiss, managed to disappear any time Gorst tried to compare the Lark biplanes against the M-1 with its greater speed, payload capacity and mail/cargo space."

Thompson Products

Famous Thompson Products calendar by artist Charles H. Hubbell featured Pacific Air Transport M-1

While in Washington, D.C., the senator from Gorst's home state of Oregon arranged for Gorst and Alexander to see President Calvin Coolidge, under whose administration the Kelly Airmail Act had been passed the previous year.

The key money man in the Pacific Air Transport organization was not Gorst but Claude Nichols Comstock who accompanied Gorst and Alexander on the trip around the country. Actually Claude Ryan had 'found' Comstock for Gorst, who badly needed additional financing.

Landing in Los Angeles one day at the end of a scheduled airline trip he had piloted from San Diego, Claude saw a chauffeur-driven Pierce Arrow drive up and a well-dressed businessman in a fur collar coat step out.

"I'm C. N. Comstock and I read in the Los Angeles Times recently about the new airmail contracts. Do you suppose the company trying to get the route on the Pacific Coast is open for some additional financial backing?"

Obviously this was a ready-made opportunity for Ryan. It might make possible consummating the deal with Gorst. Subsequently Claude arranged a meeting between Comstock, Gorst and Alexander.

Following the nationwide tour looking for planes suitable for the airmail run, Alexander returned and reported to Claude, "I've managed to keep them, so far, from buying anyone else's airplane. At times I've had to resort to a bit of doubletalk so they wouldn't make a commitment with other airplane manufacturers and they're a bit mad at me. So from here on out, Claude, you had better deal with them."

Illustrated Daily News

First airmail from Seattle is handed to Los Angeles postmaster by C. N. Comstock of PAT. Pilot Claude Ryan at left

It appears that Comstock, by then vice president of the airline, had raised the question as to who was going to put up the money for the airline operation. He said he was willing to advance $32,500 if others would carry their share of the load. In the course of a conversation with Alexander, Comstock said that if Claude Ryan was so convinced his plane was best, then the Ryan company ought to be willing to invest in the airline to prove how good the M-1 was.

Feeling himself cornered, Alexander said, "Put me down for $15,000." When Alexander called this to Ryan's attention on his return from the tour, Claude made it clear in no uncertain terms that his objective in business was to get money from potential airline operators, not give it to them to finance their operations. Since Alexander didn't have the $15,000, Ryan said that "J.B." was going to have to get out of the squeeze as best he could. Perhaps he ought to admit that he simply wasn't authorize to commit the company. (An early prospectus of Pacific Air Transport lists Ryan Airlines, San Diego, as subscribing $16,500.)

Thus, Alexander had to ask to be counted out, but obviously this hurt Ryan's chances to sell the M-1s. However, Claude went to Gorst and said that it was necessary to get the order down on paper promptly if he wanted to be qualified to start the new airline operation.

Finally Gorst, Ryan and Comstock came to an agreement. Ryan had talked with Comstock to try to smooth the problem over and said he would be willing to make some additional tests on the plane, particularly as related to the Post Office's requirement to carry 600 lbs. of mail. Sandbags were obtained, filled and weighed and a 600 lb. load placed in the front cockpit. Ryan was completely confident of his new Whirlwind-powered plane and wasn't surprised that the M-1 more than did the job in carrying the load.

At this point, Gorst — always an aviation enthusiast — got re-energized and as a result of this meeting in Los Angeles, put up the deposit on the first six planes (later a seventh was purchased). As part of the final settlement in getting the order, Claude also agreed that he himself would make a survey flight with the Whirlwind-powered M-1 between Los Angeles and Seattle to prove the plane's capability.

Only a month after the first flight of his new monoplane, Claude left on the pioneering M-1 flight for Pacific Air Transport, carrying a token amount of mail between each town, where he and his flying companion, Gorst—and sometimes Comstock — were royally welcomed by delegations headed by the local postmasters.

First Ryan M-1 with Whirlwind J-4 engine gets a final check at Dutch Flats before leaving on survey flight

Ryan Aeronautical Library

The 120-mile flight to Los Angeles was made in 62 minutes, a good omen. Flight records were set between other stops — Bakersfield, Fresno, San Francisco, Medford, Portland and Seattle. The arrival of the M-1 in Seattle on March 18, 1926 received excellent coverage in the newspapers.

The M-1 was living up to Ryan's belief that what this country's airmail lines needed was a monoplane of clean design to replace the awkward, slow biplanes of the era. The new Ryan commercial plane set a mark of 7 hours, 3 minutes flying time between San Francisco and Seattle, slashing by almost an hour the record set the previous year from Seattle to San Francisco by an Army deHavilland.

But other laurels were to be added by the M-1. On the return flight, the plane stopped over the weekend at Vancouver, Washington, while Gorst visited his family in nearby Coos Bay, Oregon. Ryan remained at Vancouver, and re-

George Allen in an M-1 had it out with an Army pursuit plane and won

newed acquaintances with some of the Army fliers he had known while in the service. Among them was Lt. Oakley G. Kelly, who had won fame with Lt. John A. Macready for making the first non-stop transcontinental flight. At the Vancouver Army Base was Kelly's two-place, 400 h.p. deHavilland, considered one of the fastest and most powerful in the Air Service.

The Army fliers looked with jaundiced eye on Ryan's M-1. Ryan's remark that it would fly 135 miles an hour evoked considerable skepticism. When Kelly's deHavilland attained that speed, the Air Service had exhibited great pride. Flush with the triumphant receptions on the survey flight, Ryan confidently predicted he could beat the powerful DH.

That was too much for the Air Service men to swallow. They challenged Ryan to make good his claim. A race was arranged, and duly publicized. Air Service mechanics swarmed all over the deHavilland the night before, tuning it up. Special fuel was poured into its tanks. Meanwhile, the M-1 sat to one side, forlorn and apparently neglected. Ryan had no mechanic and the only fuel he planned to use was ordinary automobile gasoline.

From Portland, just across the Columbia River, a crowd surged into Vancouver to witness the race, which the newspapers had touted as one between the Army's pet plane and a small commercial craft being demonstrated for airmail use — 400 h.p. versus 200 h.p.

Over a triangular course in which a column of the Interstate Bridge constituted one pylon and two other prominent points provided the other turning markers, the easily maneuverable M-1 sped past the deHavilland to an astounding victory.

Oakley Kelly took the occasion to comment on the government's failure to keep abreast of commercial aviation development. In an interview with the Portland Oregonian, Kelly explained, "Here's my machine with a 400 horsepower motor with less speed than Ryan's with a 200 horsepower motor. My machine was built by the government in 1917, and Ryan's was built this year. I'm dragging around a lot of struts and wires that aren't necessary to the modern plane. In other words, the Army needs some new machines to keep up the pace with private and commercial owners."

Ryan had proven his point. And while he won the battle, he almost lost the war, as Claude later explained.

"On my way back down the coast, I went into Crissy Field in San Francisco. I was still a reserve officer—Lieutenant, with pursuit pilot rating — and the General there came up to me and said, 'Young man, you have greatly embarrased the Army and you'll live to regret it."

"And I think I did because it was years before I ever got a government contract!"

The M-1 not only bested the DH but, after returning to San Diego, also outflew the lastest Army Air Service pursuit plane.

With Ryan Airlines' pilot George Allen at the controls, the M-1, powered by the 200 h.p. radial engine, was able to turn inside the Boeing-built Thomas-Morse MB-3A pursuit plane in combat flight practice, and generally outmaneuver it. From a standing start the monoplane was able to get into the air in less than half the distance the MB-3A did and climbed to 2300 feet while Lt. Ned Schramm was getting the TM purusit plane up to 2000 feet.

H. J. van der Linde

Colorado Airways purchased Hisso-powered M-1s for Contract Air Mail Route No. 12

During the 2600-mile airline survey flight—in and out of all kinds of fields under the varying and often severe weather conditions of early spring — the M-1 had averaged a cruising speed of 114.08 miles per hour. Its advertised cruising speed was 115 miles per hour.

The successful survey trip to Seattle and return, and competitive flights against Army planes, established the M-1 as a solid contender in the still small air transport field. In all, seven planes were delivered to Pacific Air Transport, which opened service on Contract Air Mail route No. 8 on September 15, 1926.

Colorado Airways of Denver, operator of C.A.M. No. 12 between Cheyenne, Wyoming and Pueblo, Colorado, purchased a number of Hisso-powered M-1's. Other airlines, including Clifford Ball, Inc. (Cleveland-Pittsburgh) placed orders, as did some individuals and aircraft dealers. Twenty-three planes were sold the first year of production. The M-1 became the first volume-produced monoplane in the United States. Expansion was necessary. Ryan and his partner, Mahoney solved the problem by renting an abandoned fish cannery building in San Diego to expand production.

The M-1 continued to pile up notable flight records.

Flying one of the first M-1's on a May test flight preliminary to the opening of the Pacific Coast mail run, Lee Schoenhair, one of the PAT pilots, made the 1050 mile Portland-to-Los Angeles journey nonstop in nine hours twenty minutes.

Newspapers and wire services sought the help of Pacific Air Transport in chartering M-1's to speed up coverage of important news events—and made news themselves in so doing.

An M-1 gets serviced at Vancouver, Washington. Note PAT insignia

Ray Lindsay/Vic Stuhr

When the dirigible *Norge* flew across the North Pole on May 11, 1926 and continued on another 2200 miles to land near Nome, Alaska, photos were sent by surface to Seattle. George Allen, just hired away from Ryan Airlines by Pacific Air Transport, flew there from San Diego in the third of the M-1's to pick up the pictures and speed them to news centers along the Pacific Coast.

"North of San Francisco," he wrote, "we ran into heavy storms and considerable fog, conditions which showed us what we might expect in the year-round operation; what we had to look forward to when we started regular schedules later in the year."

Things also began to happen fast the July morning the radio news flashed word that "kidnapped' evangelist Aimee Semple McPherson had been found on the desert near Douglas, Arizona.

Vern Gorst had been contacted in PAT's San Francisco office to send an M-1 to Arizona to pick up pictures and get back to Los Angeles as fast as possible. Gorst called San Diego and talked to Claude Ryan who had just learned the news from Max Miller, a reporter for the San Diego Sun. Would Claude take one of the M-1's waiting delivery to PAT and fly the charter himself? He would.

After a fuel stop at Tucson, Claude sped on to Douglas and was the first plane to arrive. While waiting for the photographs, Major C. C. Moseley of Western Air Express arrived in a Liberty-powered Douglas mail plane as did Boyd Monteith Shelton, a former Ryan Airline pilot, in a DH biplane.

Four other planes started for Douglas but never arrived. Finally the pictures showed up at six in the evening. By the time Claude reached Tucson to refuel on the return trip it was dark but he continued on through the night until turned back by fog. After a few hours wait at March Field, Riverside, the M-1 took off again but the fog in Los Angeles had still not raised, so Ryan finally landed in a barley field on the clear side of one of the foothill passes at the north end of the San Fernando Valley.

Police cars sped the pictures into Los Angeles and extra editions were soon on the streets. George Allen, arriving at the field by auto, climbed into the M-1 and took it on into San Francisco, scoring a news scoop there also. Of the two other planes at Douglas only one took off and it had to remain overnight at Yuma.

Ryan and the M-1 has scored important points once again.

* * *

United Air Lines

Art Starbuck, PAT pilot, center brings his M-1 into Angelus Mesa Field, Los Angeles

By mid-summer 1926, the manufacturing side of Ryan Airlines' operations was beginning to edge out the airline and local flying operation in importance, and the need for a full-time engineer was becoming pressing.

The Waterhouse and Royer arrangement had been unsatisfactory; Jack Northrop and Art Mankey had done a good job, but their services were only part-time and temporary.

On the east coast, one engineer who was gaining stature, particularly in Claude Ryan's eyes, was Guiseppe M. Bellanca, designer for Wright Aeronautical Corp. of the Wright-Bellanca six-seater monoplane which the engine manufacturing company was using as a test plane to demonstrate its new Whirlwind engine.

We do not know the correspondence which went on between Ryan and Bellanca before Claude's letter of August 27, 1926 to the eastern designer, but clearly Ryan hoped the two might join forces.

"Our M-1", Claude wrote, "is the same general type of externally-braced high-wing monoplane as yours except smaller, being three-place. It was designed with the idea of quantity manufacture and low selling price. We have not sacrificed material or workmanship, and the safety factor of our plane is high. We intend to perfect one or two designs and standardize on their production.

"Our flying division operates the Los Angeles-San Diego Air Line, flying fields at both terminals and a school of aviation.

"The aircraft manufacturing company is located separately in a factory on the bay front some ten blocks from the center of San Diego. This concrete building (the abandoned fish cannery) has 13,400 square feet on two floors with a spur track of the Santa Fe Railway alongside. It is located within a block of the proposed field (later Lindbergh Field). The hangars and general office are now located at our San Diego airport several miles from the factory.

"The first M-1 monoplane was flown in February and since then we have built fourteen planes. Our rate of production is about three planes per month, with orders coming in much faster. We feel very much encouraged over our outlook.

"Financially the company is sound. We have sufficient capital for our present needs and should we desire more it is available.

"To make a perfectly rounded out airplane manufacturing organization I can think of no selection we might make which would mean more to the company than the affiliation of an engineer of your ability.

"We would be interested to know the nature of your arrangement with the Wright Company or your idea of a possible working agreement that would prove mutually advantageous."

But nothing came of Ryan's overture to Bellanca.

Hudek/Ford Museum

Ryan M-1 was flown by Vance Breese in 1926 Ford National Air Tour

Selling the M-1

WITH the Pacific Air Transport order tied down and the first planes already in production, Ryan and his partner turned their attention to obtaining follow-on business.

"Monoplanes still have a bad reputation," Claude told Mahoney, "and we've got to prove that the wings of the M-1 are strong enough. So we ought to enter all the air meets we can and go on the Ford Reliability Tour. We need to demonstrate acrobatics at every opportunity in order to satisfy the skeptics.

"I can't take the time off to go myself, but we should hire a good acrobatic pilot to give these demonstrations."

Ryan tried to get Lt. Ned Schramm, who had been his old Army instructor at March Field – then stationed at Rockwell Field, San Diego – but Schramm couldn't get the necessary leave of absence.

Some time earlier, while in San Francisco on a charter flight, Claude had seen a Sunday air show where an obviously skilled pilot flying a Thomas-Morse, was holding the crowd's attention. His name was Vance Breese, a youngster Ryan's friend Sam Pursell had recently taught to fly.

Later, in trying to find a good demonstration pilot, Ryan contacted Pursell and asked more about Breese. Pursell told him that Vance "will fly just for his expenses if you haven't money to pay him; he just loves to fly." The outcome was that Breese was engaged to fly one of the PAT M-1's in the Mile High Air Meet at Denver early in August, where he won three first prizes, including the speed and altitude contests, as well as receiving the Grand Sweepstakes Trophy for the best all-round performance.

After Denver, Breese and J. B. Alexander went on to join the Ford Reliability Tour later in August. Everywhere Vance did a tremendous job of acrobatics, gaining a fine reputation for himself as a pilot and for the M-1 as an outstanding airplane capable of the most severe maneuvers.

This, of course, validated the fact that the M-1 was safe, and that here was one monoplane which could really "take it".

Alexander recalls the days of this, the second Ford Tour. "Vance would dive down, then pull up sharply until he was just above a stall, standing the M-1 on its tail. He'd then tip over, cut the gun and glide down into a perfect stall landing in the middle of the field.

The great galvanized Loving Cup is awarded twice a year, at the Nats and the Trade Show. Art Laneau of Ambroid, was the lucky recipient at Chicago and Cdr. Paul Boyer, assisted by John Worth, your editor, music and color guard made Art realize that he was a worthy choice for the Cup.

SCENE: At The '65 Trade Show

By M.A.N. at WORK . . 1965, YEAR OF THE AWARDS AT HOBBY INDUSTRY OF AMERICA TRADE SHOW. PICTURES ON THIS PAGE DEVOTED TO THE MANY PRESENTATIONS.

Matty Sullivan, on right, of Sullivan Products and one of modeling's best friends, receives his Fellowship Award from Academy's John Worth.

Somehow, your editor managed to be in four of the five pictures in this group. Here, he is receiving his AMA Fellowship Award from John Worth.

Tom Sutor, left, presents the HIAA with an award of appreciation for the Industry's continuing support of the King Orange International contest.

The Model Aeronautics Division's guest of honor at the trade show and luncheon, Adm. George P. Koch and M.A.N.'s Publisher, Jay P. Cleveland.

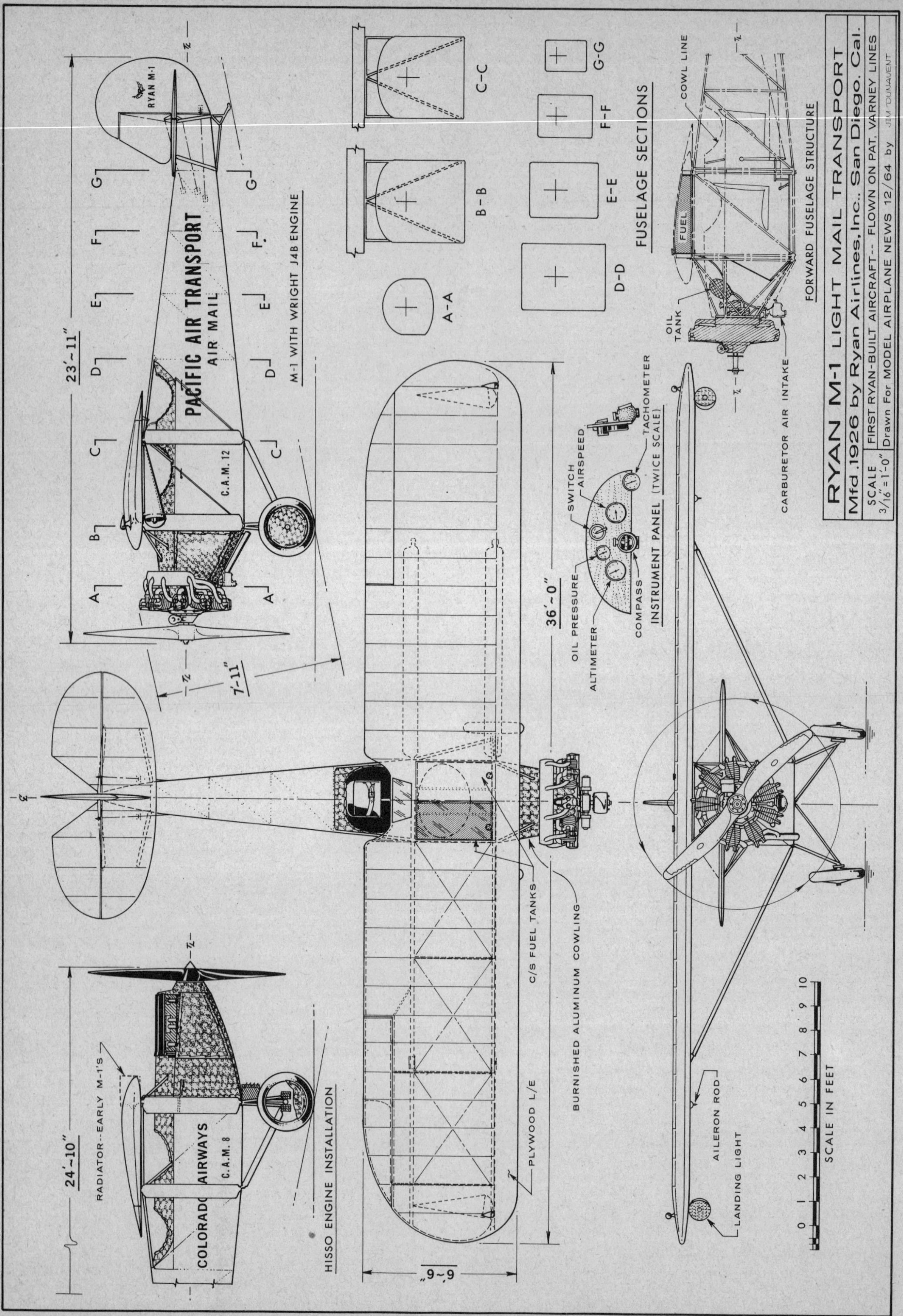
RYAN M-1 LIGHT MAIL TRANSPORT
Mfd. 1926 by Ryan Airlines, Inc., San Diego, Cal.
FIRST RYAN-BUILT AIRCRAFT-- FLOWN ON PAT. VARNEY LINES
SCALE 3/16" = 1'-0"
Drawn For MODEL AIRPLANE NEWS 12/64 by
JIM DUNAVENT
M-1 WITH WRIGHT J4B ENGINE
PACIFIC AIR TRANSPORT
AIR MAIL
C.A.M. 12
RYAN M-1
23'-11"
7'-11"
36'-0"
6'-9"
24'-10"
FUSELAGE SECTIONS
A-A
B-B
C-C
D-D
E-E
F-F
G-G
FORWARD FUSELAGE STRUCTURE
COWL LINE
FUEL
OIL TANK
CARBURETOR AIR INTAKE
INSTRUMENT PANEL (TWICE SCALE)
SWITCH
AIRSPEED
TACHOMETER
OIL PRESSURE
ALTIMETER
COMPASS
C/S FUEL TANKS
BURNISHED ALUMINUM COWLING
PLYWOOD L/E
AILERON ROD
LANDING LIGHT
SCALE IN FEET
RADIATOR--EARLY M-1'S
COLORADO AIRWAYS
C.A.M. 8
HISSO ENGINE INSTALLATION

RYAN M-1 PERFORMANCE	Wright Whirlwind	Hispano Suiza
Power rating	200 h.p.	150 h.p.
Cruising speed	115 mph	110 mph
High speed	135 mph	125 mph
Landing speed	45 mph	45 mph
Cruising range	500 miles	400 miles
Initial climb	1,200 fpm	1,000 fpm
Climb in 10 min.	9,000 ft.	7,000 ft.
Service ceiling	17,500 ft.	15,000 ft.
Payload	600 lbs.	500 lbs.
Weight empty	1,600 lbs.	1,550 lbs.
Useful load	1,200 lbs.	1,150 lbs.
Gross weight	2,800 lbs.	2,700 lbs.
Over-all span	36 ft.	36 ft.
Over-all length	24 ft.	24 ft.

"All of the old timers would hold their breath every time Vance did this. Finally a committee consisting of R. W. (Shorty) Schroeder and Eddie Stinson came over to me and complained.

"They said, 'J.B., tell that kid to cut out that stunt, because if that engine ever quits at the top of that stall, he's coming in tail first and ruin the whole tour.'

"The very next time, Vance cut the throttle right at the top of the stall and we sold some airplanes!"

In the formal part of the tour Breese and the M-1 placed eighth. But for spontaneous demonstrations Vance had no peer.

In a demonstration at Santa Monica after the tour, Ryan saw the type of flying that Breese had been doing. One particularly effective stunt was to dive the plane toward the field, then roll over 90 degrees until the wings were vertical and fly across the airport, getting enough lift from the side of the fuselage to maintain altitude.

Claude had to admit that while "Vance took chances I would not have approved of," the reputation of the M-1 as a rugged plane had certainly been enhanced.

On their return to San Diego, "J. B." stated that the M-1 had flown 9000 miles at an average of ten miles to the gallon of gas, and 3000 miles to each gallon of oil. Except for gas and oil the only expenditure had been $1.60 for an inner tube!

* * *

Claude Ryan was his own best salesman; no one else was quite as adroit in coining slogans like "the plane that pays a profit", or in writing advertising copy. After all it was Claude's baby and he wanted it placed in the best possible light.

As a pilot, he knew what would interest other pilots. This June 1926 ad, for example, which featured the M-1's capability in stunt flying —

"Loop it; barrel roll it; dive it a thousand feet and pull back quickly on the stick. Try to whip-stall it; try to spin it.

Tom Mathews

The Ryan sales force; from left, T. Claude Ryan, J. B. Alexander, B. F. Mahoney

Tom Mathews Frank Wiley

Vance Breese, beside cockpit, turned the M-1 inside out demonstrating its excellent aerobatic qualities

Ryan Aeronautical Library

Claude Ryan . . . his own best salesman, with Super-Rhone-powered M-1

Wind it up in a tight spiral, full throttle, kick top rudder, give it everything, then float it down and look it over.

"A biplane would probably need re-rigging. But the Ryan M-1 will be just as tight as the day it left the factory.

"A monoplane can be built lighter, faster and more efficient per horsepower than a similar biplane. The monoplane is aerodynamically correct. Perhaps you didn't know that it could be built stronger and safer than any biplane now on the market. The Ryan M-1 is an example.

"Distinctly superior, with a price range from $2890 to $8400." (One price list quoted $3750 for the Hisso M-1 and $8100 for the Whirlwind M-1 with standard equipment.)

By August, *Aero Digest* magazine was carrying a special 8-page advertising section on the success of the M-1.

Many a tall tale of the early M-1 days has been told by those who shared the series of often bizarre incidents, many of which are described in "Ryan, the Aviator" (McGraw Hill Book Co.)

After the first group of six planes for Pacific Air Transport had been built, Ryan and Mahoney kept on producing more M-1's, expecting that orders would continue to come in, but soon they were in debt to the bank financing the work-in-process in the factory.

With five unsold planes in their inventory, Ryan was happily surprised one day by a caller who wanted to buy all of them for a big promotion stunt. A week later Claude had the contract and down payment in hand.

Ryan's euphoria was short lived. The Hisso-powered M-1's, it turned out, were being purchased by backers of a revolution in Mexico and were impounded when the United States Government got wind of the impending sale.

Finally, with legal help, Ryan got the U.S. Attorney to release four of the planes; the fifth would be held for evidence and was locked up across from the Ryan Airport in the Marine Corps Depot. The four planes were soon sold, enabling the company to continue paying its bills.

But it was many months before the fifth plane was returned, and then it was in such poor condition it had to be virtually rebuilt. It then went to the Menasco Motors Co. in Los Angeles which used it as a test bed for their conversion of a French Salmson engine. Later, in 1927, George Allen was flying the same "Mexican Revolution" plane for Northwest Airways of Montana.

Two other Ryans — an M-1 and an M-2 — also showed up in Montana. Frank Wiley, one of the early Ryan Airlines' pilots, took delivery of Company Serial No. (c/n) 23 in San Diego on March 26, 1927 for the Westland Oil Company of Scobey, Montana and

. . . five unsold planes in the inventory

Frank Wiley

H. A. Erickson

Frank Wiley

Airmail pilot Lindbergh, left photo, got to fly the Westland Oil Company M-1 whose pilot Frank Wiley poses alongside

Minot, North Dakota. Several days earlier a young airmail pilot named Lindbergh — recently arrived from St. Louis — had flown it on a test flight at Ryan's Dutch Flat airport.

(The last remaining M-1 in existence, No. 23, went on permanent display in the San Diego AeroSpace Museum in 1968.)

Two and a half months later Wiley was back in San Diego to accept delivery of an M-2 for State Senator John W. Schnitzler, banker and flying farmer of Froid, Montana.

In September 1926, Tom Mathews and Ryan pilot Joe Barrows had taken off in an M-1 for Northern California to follow up on fifty prospects who had written in for information. Claude Ryan, only half jokingly told them not to come back if they didn't sell some planes. Unfortunately most of the "prospects" were kids. The sales trip was a failure and Ryan told them to come home.

Nearing San Diego they had a forced landing. Mathews called in to tell Ryan his plane was in a barley field, undamaged but a little worse for wear and tear, and that "we'll see you later".

"And you know," Ryan later recalled, "that was the last I saw of either Tom or Joe for years!"

While the new, privately operated airmail lines were the best prospects for the M-1, a number of private owners, dealers like Von Hoffman Aircraft Co. of St. Louis and Yackey Aircraft of Chicago, and flying service operators were also also purchasers.

Typical of private owners was Elwood Lloyd, financial editor of the Los Angeles Evening Herald, who bought an M-1 early in 1927 to commute the hundred plus miles between Los Angeles and his desert home at Thousand Palms. Lloyd operated from Rogers Airport, just five miles from his downtown office.

Another buyer was Clarke Chapman of Long Beach, California, who hired Ryan Airlines' pilot Bill Monday to fly him and haul supplies to mining properties he had in the deserts of Southern California. J. F. Dexter, general contractor, of Dallas, Texas, was also an M-2 buyer.

Manitoba Basin Mining, Ltd., headed by W. H. (Hal) Wolfe of The Pas, Canada, took delivery of Serial No. 22, an M-2 powered with the 180 h.p. Hisso engine. At Toronto it was equipped with skis for operation in Northern Canada, and carried Canadian license G-CAJK. Later the engine was changed to the popular Wright Whirlwind.

The mining company also had an interest in Eastern Canadian Airways, Limited, to which the plane was later registered. Still later, in 1932, when the company pilot, Howard Watt, was the registered owner, the plane made a forced landing on an ice floe in Quebec and drifted out to sea. The plane was lost but the crew was saved.

One of the last M-1/M-2 series Ryan monoplanes built was Serial No. 28. Delivery of the Hisso-powered M-2 was taken July 17, 1927 by Walter R. Bullock for its owner in Minneapolis.

. . . to sea . . . on an ice floe

Bill Monday hauled supplies to mining properties in an M-1

Bill Monday

"The flight," wrote Bullock, "was uneventful all the way. Except for sunburn there was no discomfort or malfunctioning of the aircraft. It was the finest performing airplane I had flown.

"My passenger, who was not a pilot, gave me a few moments of concern when trying to reach into the rear cockpit and take over. Fortunately there were no front cockpit controls or I might have had real trouble."

As Bullock explained, "My passenger was an alcoholic."

* * *

A whole book should be written on the early days of Pacific Air Transport. Indeed files of the Ryan Aeronautical Library contain a very complete history of the operation, which later was purchased by Boeing Air Transport and subsequently became the Pacific Coast route of United Air Lines.

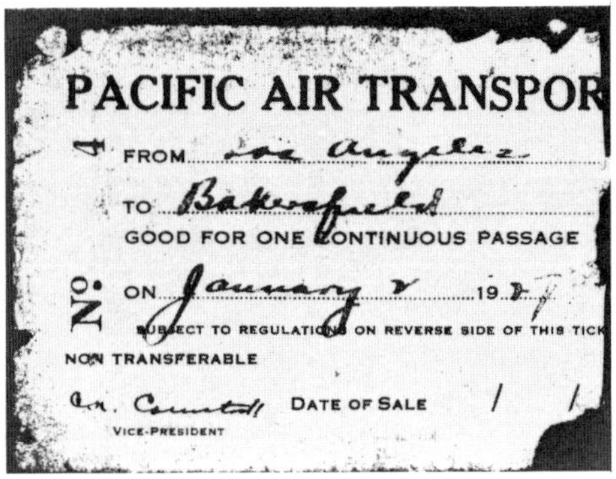

Scholer Bangs

Ticket No. 4 on Pacific Air Transport went to a Los Angeles newsman for flight to Bakersfield

Pacific Air Transport had not only the longest but perhaps the most hazardous of the new privately operated airmail routes. Among the hazards were lack of money — and skepticism.

"One of our greatest handicaps," pilot George Allen wrote, "was the fact we had very little money but plenty of work in setting up the airway. We were fighting an uphill battle. Even everyone in aviation claimed it was impossible to figure on year-round operation over a night route and to maintain any degree of regularity and certainty.

"Getting the authorities of cities scheduled to receive airmail service to agree on providing fields was also a great problem. We finally got action by notifying Chambers of Commerce that unless fields were established by a set date we would recommend to the Post Office Department that some other city along the line be chosen."

United Air Lines

"Old No. 7" in front of its hangar at Los Angeles

In February, 1927 *Aero Digest* magazine reported on operations over PAT's 1121-mile route:

"Up to November 20, 1926 the service was 98 percent perfect. Since that time, delays have occurred and two of their planes have met with serious accident. These delays and accidents resulted almost entirely through the impossibility to secure adequate knowledge of weather conditions. These obstacles now are being overcome through the receipt of accurate weather reports, and through the installation of a system of radio communication.

"With the experience of the worst season of the year behind them, a season that has been almost unprecedented for its number of days of heavy fogs, they look into the future with high hopes, having learned many things and overcome seemingly insurmountable difficulties.

"As the time schedule required their leaving Los Angeles at midnight and Seattle before daylight, the company lighted the airways between Seattle and Portland with ten revolving beacons of 7,500,000 candle power each, and twenty intermediate beacons of 400,000 candle power each.

"They have carried 19,658 pounds of mail, through the end of December, and 102 pay passengers in their open mail planes. During the winter months however, passenger carrying has not been encouraged. The fare from Seattle to Los Angeles is $132.50. Intermediate stop fares are in proportion to distance.

"They propose to install, at the earliest practicable date, the most modern cabin planes to care for the business that is inevitable as soon as it becomes generally known that the mail and passenger time between, for example, San Francisco and Portland is six hours via plane, as against twenty-seven hours via the fastest train."

Aircraft and pilot mortality the first year of operation was indeed high — too high — mainly because airway aids were inadequate and the pilots were committed to the philosophy that "the mail must go through."

Charlie Widmer had to bail out after leaving Los Angeles one midnight, but his mechanic, Don Rossiter, jumped too late or stayed with the plane and was killed when it crashed in the Hollywood Hills. Jim Rutledge had to "hit the silk" in bad weather near Bakersfield. Eddie Neher, carrying the company's accountant as passenger, crashed in a storm front west of Fresno, killing both occupants. George Allen met a line squall over the Tehachapi Mountains forcing him down to ground level. He came into Bakersfield with mesquite brush on his landing gear.

Later, over the Tehachapis, Dick Bowman had the carburetor and wings ice up so severely that he bailed out. Landing unhurt at 2 a.m., he walked out at dawn, but it was a week before the wrecked M-1 was found.

Pat Patterson failed to clear a ridge at Ashland, Oregon, one foggy morning, strewing parts of the M-1 and his cargo of mail pouches for three hundred yards up the slope. Art Starbuck, while deadheading as a passenger into Los Angeles, got involved in a crash when the pilot misjudged his sinking speed and tripped the wheels on a high-tension line, resulting in a blazing wreck.

Several times planes landed in San Francisco Bay but "fortunately", said George Allen, "all landings in the water there were made close enough to boats that crew and passengers were saved."

Forerunner of the Brougham and "Spirit of St. Louis" was the Bluebird, an enclosed M-2

H. A. Erickson

"Bluebird" and "Brougham"

BY FALL 1926, the business aspects of Ryan Airlines' manufacturing activities were taking all of Claude's time, leaving little opportunity for the engineering and manufacturing side. And Ryan was getting scant help from his partner, B. F. Mahoney, whose interests reportedly ran less to business and more to promotional ideas, having fun and living the life of a playboy.

Hawley Bowlus was concentrating on producing the planes, and J. B. Alexander on selling them. While the M-1 was off to a good start, the company was extended financially beyond good business limits.

In a news release September 18, 1926 Tom Mathews declared that "the production end of the Ryan aircraft company has been developing so rapidly, and the demand for the San Diego built ships so steadily increasing, that it has become obligatory for the concern to reorganize, increase its capital, and devote its energies to the manufacturing business."

The news article went on to say that Claude Ryan was in conference during the greater part of the week with San Diego and Los Angeles businessmen who wished to acquire an interest in the aeronautical firm.

Mathews announced that the firm was then employing 50 men in building Ryan monoplanes already contracted for and that the company's plant at the foot of Juniper Street (the former fish cannery) was taxed to capacity.

"We have contracts for the delivery of 28 planes by March 1, 1927," said Mathews. "Twelve of these ships are to go to the Yackey Aircraft Company of Chicago, 12 to the Mono Sales Company of Denver, and 4 to A.J. Thille of Ventura. In addition to these planes we are building six mail service planes for Colorado Airways for the air mail service between Denver and Cheyenne."

Sales manager J. B. Alexander's enthusiasm — and Mathews' publicity — frequently ran rampant. Many of the orders from M-1 distributors were obviously not backed by cash deposits, as explained years later by Mathews:

"You can see how Yackey could deposit $500, and then toss off a nice order for an even dozen planes, hoping some desperate contractor among the new air mail operators would take them off his hands by paying a nice bonus. It never happened."

Clearly it was time to incorporate and bring in more capital. J. B., who

H. A. Erickson

Claude Ryan's own M-2 was powered by German-built Ryan-Siemens engine

Fred G. Borsch

Rebuilt Douglas Cloudster carried passengers on Ryan's Los Angeles-San Diego Air Line

agreed, reported that "Mahoney was unavailable for conference with the lawyers and filing of the necessary papers, always taking off for a trip on his boat when most needed."

Mahoney was a likeable but difficult young sportsman who had inherited considerable money. Claude had taught him to fly early in 1925. They were a strange pair. Ryan was all business; Mahoney all play.

Wanting to have some role in aviation, Mahoney suggested to Ryan that if he'd like to expand and needed financing they could set up a partnership.

Claude figured up the value of the small fleet of planes he then owned. Mahoney matched it with $7500 and with this much working capital the partners embarked on the Los Angeles-San Diego Air Line venture. Then, out of the never-incorporated Ryan Airlines, Inc., the M-1 manufacturing business had developed.

Ryan and Mahoney had been partners only a year-and-a-half when Claude came to the realization that their business relationship would have to end. It settled down to a question of which would buy out the other. Since all Claude's assets were tied up in the business, while Franklin had other sources of income, the decision was pretty obvious.

On November 23, 1926 the partnership was dissolved. Claude sold his remaining half-interest for $25,000 and a new Ryan M-2. Six days later the Legal Notice appeared, making it official. It read, in part:

"Said T. Claude Ryan has retired from said firm and business and has sold all his interest therein and in the assets thereof to said B. Franklin Mahoney, and said B. Franklin Mahoney will continue the business at the same place and under the same firm names."

But while Claude retired from an ownership role he agreed to continue as general manager until Mahoney could find someone to run the business for him. Mahoney offered $300 a month, but Claude reminded his former partner that it was more than the business would justify, and that he'd handle it for $200 a month while helping find a new manager.

At the same time the decision was made to phase out the Los Angeles-San Diego Air Line operation in favor of concentrating on manufacture of the M-1 monoplane.

* * *

The M-1 had been a modest but nonetheless real success story. Although it had been flying for nine months, it had been preceded by several years of gestation.

The next logical step would also take time, as had the M-1. It would be a cabin plane, one which could hold four passengers, in addition to the pilot, instead of the two passengers carried in the open cockpit of the M-1.

As with the first M-1, the job of building the new cabin plane was given to Hawley Bowlus, the company's construction superintendent who had been a fellow flying school student of Claude's ten years earlier.

This time Hawley was to have some in-house engineering help. Walter O. Locke, just out of Massachusseets Institute of Technology, had arrived in San Diego early in 1926 to become a flight and mechanics student of the Ryan School. By fall, Walt became an earn-as-you-learn employee and custodian of the company's drafting board and T-square. Because of his engineering background he became "resident engineer" assisting Bowlus with the engineering calculations and stress analysis required in converting the first Ryan M-2 (with the Northrop-engineered wing) into a 5-place cabin plane.

The M-2 fuselage was somewhat modified to accommodate the four passengers, but the narrow-tread landing gear was retained. The standard 36-foot M-1 wing was also kept. The pilot's position was moved from the rear open cockpit of the M-1 to an enclosed position up front under the wing's front spar. The Wright-manufactured 200 h.p. high compression Hispano-Suiza E-2 model water-cooled V-8 engine was installed.

According to Tom Mathews, "Bowlus made some modifications of the fuselage bracing to change the wedge-shaped door on the right side of the M-2 to a single rectangular door on the new *Bluebird* with the bottom angled off on each side. Aft of what had been the pilot's cockpit in the M-2, was a baggage compartment with exterior door on the cabin plane."

Unlike the Ryan Standard cabin biplanes used on Ryan's Los Angeles-San Diego Air Line, which had open cabin windows, those of the new *Bluebird* were of plate glass, unfortunately not shatterproof.

After Claude's initial flight in the *Bluebird,* additional tests were run by Red Harrigan, chief pilot, but Ryan recalls that "I flew it more than anyone else.

"It was a nice flying airplane although a little hotter on landing speed than most of the airplanes we were then using."

Tom Mathews, the company's publicist, expressed a somewhat different private opinion years later than the one he projected to customers at the time he was promoting aircraft sales for Ryan. "The plane," he wrote, "was unstable directionally and very nose heavy with just the pilot aboard. Subsequent flights carried either one or two passengers, or weights in the baggage compartment.

"But everyone agreed on one thing. It was a lot nicer flying in a cabin, than in the open cockpit M-1."

With a potential complete new product line of "open and closed models in the passenger, mail and express type" under development, Mathews was turned loose on a promotional campaign.

Forty-five years later, he was to write—

"Naming the Brougham was one of those tactful suggestions every public relations man has to make sooner or later. We had just built the first Ryan cabin monoplane with the M-2 wing and closed-in fuselage. It had the E-2 Hisso 200, (less reliable than the 150 h.p. Hisso but we had a lot of experience with them in the Ryan Standards).

"Before the first flight and while I was getting lined up for publicity and sales letters, somebody named it *Bluebird!* I think it was Bowlus. Georgia Mathias (Ryan and Mahoney's secretary) liked the name. So did Walter Locke and Franklin's mother and before I knew what was happening Claude and Mahoney accepted the name and started using it.

"Of course I knew the name would endear itself to love-sick youth and old ladies, but it wouldn't sell any airplanes. I mentioned this to J. B. Alexander, who agreed. He said if I could come up with a good alternative he would help present the idea to Claude and B. F. Mahoney.

"I wrote up two pages and stuck them in a manila folder. The first page outlined various reasons why an airplane-automobile tie-in might be more acceptable to a large number of potential customers than simply a new airplane. I concluded that the simplist way to make such a tie-in was with a name that would give people confidence, because of an automobile association. The second page simply contained five names.

"I gave the folder to J. B. and after reading it he said, 'which one do you want them to pick?' I said, 'the third one.' A little later Mahoney called me in and said 'we have decided to call the new plane a Ryan Brougham. Get out some publicity and advertising on it right away.'"

In January 1927 the first ad appeared for the Ryan Brougham, a Modern Closed Airplane. The next month *Aero Digest* carried a photograph and the following article about the new 'five-place, high speed commercial light transport'—

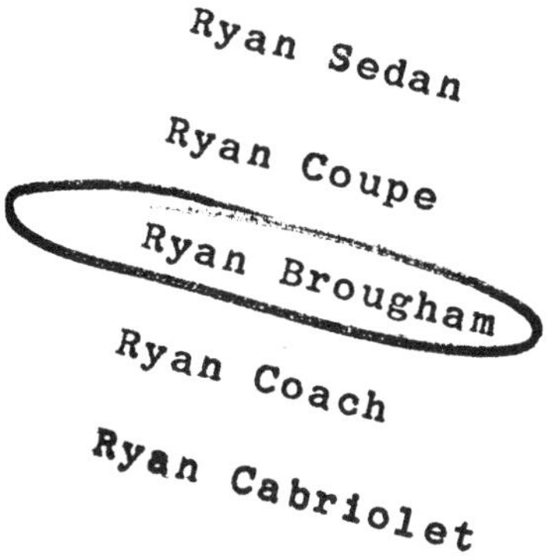
Ryan Sedan
Ryan Coupe
Ryan Brougham
Ryan Coach
Ryan Cabriolet

H. A. Erickson

First Ryan "brougham" was the Bluebird, enclosed version of M-2 with 200 h.p. Hisso-E engine

Aviation Magazine

Call it Bluebird or "brougham,' Ryan Airlines was out to sell its new cabin monoplane

Aviation Magazine

Answering the demand for a closed type plane, Ryan Airlines, builders of the Ryan M-1, announce for 1927 the Ryan Brougham, a five-place high speed commercial light transport.

The original demonstrator has for its power plant the Wright E-2 motor which gives this plane the same performance as the Ryan M-1 open model with the Wright J-4. A top speed of over 130 mph, cruising speed just under 115 mph with a landing speed of 45 mph.

Finished in Saftee-glass throughout, with all of the windows opening including the front ones, the vision for pilot and all of the passengers is excellent. Mr. Ryan tested the first plane after the sun had set, making this first landing in semi-darkness with ease. Pilots who have flown the new model marvel at its remarkable stability, ease of handling and excellent visibility.

The interior is upholstered in mohair, which tends to shut out the sound of the motor allowing the passengers to carry on a conversation in flight. The large cargo compartment to the rear of the cabin is accessible from both the outside and the inside. It is designed to hold 150 pounds of mail or express.

In other appearances the new Broughams have much the same lines as the popular Ryan M-1 open models, having the same wing spread and fuselage length.

The list prices appearing on the back of the *Bluebird/Brougham* promotional postcard with various engine installations were:

Wright Whirlwind (200 hp)	$9700
Wright E-2 Hisso (200 hp)	$5950
Hisso Model E (180 hp)	$5100
Model A Hisso (150 hp)	$4900

Even while the five-place cabin plane was being promoted, Ryan was still offering the three-place open cockpit M-1 and M-2 in paid advertisements as late as July 1927. They were priced at $4,000 for the Hisso E model and $8,100 with the Wright Whirlwind.

Ryan Aeronautical Library

The first true Brougham on its first takeoff

Debut of the B-1 Brougham

IT HAD been apparent almost from the start that with a little more from-the-ground-up engineering, the Ryan crew could come up with a better cabin plane than the *Bluebird* type converted from the M-2 open-cockpit airframe.

Several people began making simple three-view drawings of the kind of airplane they were all talking about. One was Ed Morrow, a skilled mechanic and good draftsman, who had worked with Ryan and Bowlus since converting the first Douglas plane, the open-cockpit Cloudster, to a 12-passenger cabin plane for Ryan's Los Angeles-San Diego Air Line.

Morrow's sketch resembled the *Bluebird,* being a somewhat scaled-up version of its basic design features. The wing span would be increased from 36 to 42 feet; the fuselage would be wider and longer. It would have a 200 h.p. engine – either the water-cooled Hisso E or the same Wright J-4 radial as in the Pacific Air Transport mail planes. The engine mount would be moved forward ten inches. And under consideration was a possible new landing gear arrangement with wider tread.

What the company needed was a buyer whose purchase order for the first completely new Brougham design would finance its construction. Such a prospect was Richard T. Robinson, Jr., a successful San Diego businessman and investor who had built the El Cortez Hotel, the Athletic Club and Orpheum Theater.

With Dick Robinson's order in hand, Hawley Bowlus and the small Ryan crew early in 1927 began work on the first B-1 model Brougham. The next to last M-2 was then being built. It was for a Maurice A. Warner of Minneapolis and had Company Serial No. (c/n) 28. The new B-1 monoplane would be Serial No. 29.

Walter R. Bullock

One of the last M-2s. It was constructor's number (c/n) 28

Although Bowlus, Morrow and the other Ryan shop people were practical craftsmen, the company still lacked a full-time engineer.

One day, after Claude had returned from a charter flight, Mahoney introduced him to a graduate engineer from the Pratt Institute of New York. They agreed to hire him.

Thus Donald A. Hall, who had earlier worked for the Curtiss Aeroplane Co. and then under Art Mankey at Douglas, joined Ryan Airlines on January 31, 1927.

"I did some preliminary general and structural design work on Dick

Donald A. Hall

Donald A. Hall worked tirelessly to adapt existing design to Lindbergh's requirements

Robinson's B-1 airframe" Hall recalled. "This was in the interval between February 7th and 24th."

The B-1 engineering and shop work for Robinson's B-1 c/n 29 was stopped when Ryan Airlines routines were upset by a rush order from a young airmail pilot from St. Louis who had arrived in San Diego February 23, 1927 to order a special version of the Ryan airmail plane built for a New York-to-Paris flight.

All else was dropped as the plant worked at a feverish pitch to complete the *Spirit of St. Louis* in only sixty days for Capt. Charles A. Lindbergh. That story has been so overworked that to repeat it at any length here would defeat the purpose of telling the history of the rest of the remarkable Ryan Brougham aircraft . . . But some background is necessary.

The B-1 under construction for Dick Robinson served as a starting point for Lindbergh's plane. Mel Anderson had already welded up the first Brougham fuselage. Why not take advantage of all the experience Ryan Airlines had accumulated up to that point?

Three things were clear as Hall tackled the job of adapting the existing design to Lindbergh's special requirements.

Greater wing area would be required to carry the unusually heavy fuel load. So the wing, extended from the M-2 and *Bluebird's* 36 feet to 42 feet for the B-1 Brougham, would have to be increased another 4 feet. Thus wing span of the NYP (New York-Paris) plane was 46 feet.

The engine must be the new 223 h.p. Wright Whirlwind model J-5C specified by Lindbergh and "super inspected" for the flight.

H. J. van der Linde

M-2, above; "Spirit," below, show contrast in landing gear arrangement

H. A. Erickson

And the landing gear would have to be beefed up to absorb the pounding the 2150-pound airframe would take as the *Spirit* struggled into the air with 2600 pounds (425 gallons) of gasoline for the long overwater flight.

The narrow-tread landing gear of the M-1 had been retained in the M-2 and *Bluebird.* Landing impact loads were absorbed by struts carried to the lower fuselage structure.

Perhaps influenced by a design imported from Holland in the Fokker Universal high-wing monoplane, Hall came up with a wider tread landing gear which featured a vertical member carrying loads from the wheels through struts to the wing and upper fuselage. It also carried loads to the lower fuselage as in the basic M-1 design.

The 46-foot wing is fitted to the "Spirit." M is Mahoney, X is Lindbergh, Y is "Wrong Way" Corrigan

H. A. Erickson

When work resumed in mid-May on Robinson's B-1 *GoldBug* — so named because of the paint scheme he had specified — the NYP-type outrigger landing gear for the *Spirit* was incorporated in the B-1 design, and would be retained through the Brougham series.

But Robinson didn't get his plane.

* * *

When the unknown Lindbergh made his sensational one-stop San Diego-New York flight on May 10-11, he suddenly became the most talked of pilot among those poised at Roosevelt Field seeking the Orteig Prize for the first New York-Paris flight.

The prospect that the Ryan NYP monoplane might well be the first across the Atlantic was enough motive for Mahoney, and later Hall, to go to New York to get in on the action. Then, Mahoney went on to Paris to try and join Lindbergh.

That left a vacuum in San Diego. Again Bowlus had to take over and finish up the prototype B-1, improvising somewhat in the fine details. Because of the huge gas tanks in the *Spirit* and Lindbergh's position behind them, the engine mounting and pilot's position in the still experimental *GoldBug* had to be different than in the NYP.

And no decision had yet been made on selection of the engine for the prototype B-1. Meanwhile another B-1 had been ordered, this one by William (Billy) Brock for a customer in Montana. Brock became famous later that year for an 18-day Newfoundland-Tokyo flight.

Although the Wright Whirlwind engine, especially after Lindbergh's flight, was acknowledged to be by far the best in its power class, a new Ryan Brougham with war-surplus Hisso engine could be ordered for lIttle more than half the cost of the Whirlwind-powered B-1. (The 180 Hisso engine was priced at $1100, the 200 h.p. Wright at $5050. The B-1 airframe was $4620.)

Thus, the second B-1, for Brock, was ordered with a Hisso engine. It was flown before the *Gold Bug,* using an A model Hisso of 150 h.p. It reportedly developed engine trouble and was returned to the factory. The engine was rnplaced with a 180 h.p. Hisso E and the plane was put back on the flight line as a standby for passenger and demonstration flights.

Another Hisso-powered Brougham, c/n 31 and designated as a B-2, flew with license 3007. The third of the Hisso B-1 Broughams, c/n 46 with license NC 3226, was christened *The Betty Rogers* and operated out of Rogers Airport, Los Angeles. Later it was used in the movies as *Spirit of Youth*. However, the Whirlwind J-5 radial engine, not the Hisso, became standard for the Broughams. (To this day no one has come up with a satisfactory explanation of the difference between a B-1 and a B-2 Brougham.)

What type of landing gear would have evolved for the Brougham had it not been for the heavy load requirements of Lindbergh's NYP? The first three or four Broughams, like the *Spirit,* had "trombone" type rubber shock absorbers covered by a large streamlined fairing; later all had air-oil hydraulic shocks featuring "Aerol Struts".

After Lindbergh's landing in Paris, Frank M. Hawks, a popular, nationally known pilot, wanted to get his hands on the nearest thing to the *Spirit* to fly in the annual Ford Reliability Tour, which had been started two years earlier.

Coming to San Diego, Hawks succeeded in getting Robinson to give up the partially completed *Gold Bug* and let Hawks, then flying for Texaco, have it instead. Texaco arranged for a J-5 220 h.p. Whirlwind super-inspected engine like Lindbergh's and it was installed in the B-1.

Wilhelm Schult

This Hisso-powered B-1 with its burnished aluminum panels and fairings looks as solid as a battleship

H. A. Erickson

"Spirit" fuselage rolled outside. Left to right, O. R. McNeel, Douglas (Wrong Way) Corrigan, Bert Tindale and Hawley Bowlus

New strut-shock absorber system is tried out on the ninth Brougham built

H. A. Erickson \Hugh Belknapp

H. A. Erickson

Frank Hawks' "Goldbug," the first Ryan Brougham built

In making the initial test flight in the Hisso-powered Brougham, Red Harrigan found some difficulty because of a nose-heavy condition. At about 3000 feet altitude he began a wide, gentle turn and came back in to land on power, fanning the tail down to keep from nosing over. The engine had been mounted too far forward. The front end was redesigned and the pilot's seat was moved nearly a foot aft.

With Hawks as owner, the former *Gold Bug* Brougham, declared "manageable" by Harrigan after minor design changes, was christened *Pride of San Diego* and given license 3009. Landing lights, which had been faired into the leading edge at the wing tips, reportedly interferred with aileron control, making for a rough ride. The sales order, for $8250, was dated May 28, 1927.

Bunnell Photo

Frank Hawks and his bride

Hudek/Ford Museum

. . . in Ford Reliability Tour

Bunnell Photo

. . . as "Pride of San Diego"

Ryan Aeronautical Library

. . . as "Miss Maxwell House"

Just a month after Lindbergh left San Diego in the *Spirit,* Hawks, accompanied by his bride and by D. W. Campbell of the San Diego Chamber of Commerce, took off for St. Louis. From there they flew the *Gold Bug* on to Washington, D. C., where the sister-ship of the *Spirit* flew out to greet Lindbergh triumphantly returning on the Cruiser Memphis.

In the Third Ford (National Reliability) Tour, from late June to mid-July 1927, Hawks placed eighth in *Gold Bug.* In the Air Transport Trophy competition at the Air Races in Spokane, Washington in September, E. W. (Pop) Cleveland placed eighth in his NC 1159 (c/n 38) Brougham in the New York-Spokane Air Derby. Hawks placed first in speed and third in efficiency. In the Fall Hawks made a deal with Maxwell House Coffee and the former *Gold Bug* and *Pride of San Diego* was rechristened *Miss Maxwell House* and relicensed as 1105.

Summer of 1927 was a period of many reckless, poorly conceived flights attempting to emulate Lindbergh's record. But some — like the Oakland-Honolulu nonstop flight by Army Lieutenants Lester J. Maitland and Albert F. Hegenberger, which required precise navigation to hit their destination—were more productive.

Flying in the opposite direction, west to east, there was the whole North American mainland — a fairly easy target to hit. One who tried was movie stunt pilot Dick Grace who had bought a surplus Ryan M-1, probably from Pacific Air Transport, which he had shipped to Hawaii. Grace took off in darkness from Barking Sands beach on the Island of Kauai July 4th but only covered a few hundred miles when mechanical trouble forced him to return. His crash landing on the beach virtually destroyed the plane, but Grace was uninjured.

* * *

Even as the first few B-1 Broughams were being built, the short career of their predecessor, the *Bluebird,* was coming to an end.

A. J. Edwards, race driver, auto dealer and aviation enthusiast from Muskogee, Oklahoma, had joined the Mahoney organization in February 1927 as sales manager succeeding J. B. Alexander. Claude Ryan had met Edwards on eastern trips and had introduced him to Mahoney who was looking for new management talent.

The day that Lindbergh landed in Paris, Edwards and company pilot Doug Kelley, the latter at the controls of the *Bluebird,* had taken off for Los Angeles. They had great news to celebrate and reportedly were in high spirits.

As they approached Rogers Field, Edwards opened the door in flight, to wave to someone on the ground. That spoiled the plane's aerodynamics. Kelley could hardly figure out why the plane flew so erratically as he tried to side-slip it in to a landing. In fact, he wrapped the *Bluebird* up in a ball, the plane going over on its back. Kelley was unhurt but Edwards ended up in the hospital with a fractured skull and scalp wounds from flying glass.

The plane was trucked back to San Diego and rebuilt to its original M-2 configuration. They hardly needed the *Bluebird* as a cabin plane prototype since the first of the larger, more modern B-1 Broughams was nearing completion.

Although Ryan built only the one enclosed *Bluebird* — modified from an M-2 airframe — the cabin idea had some merit for other M-1s and M-2s, particularly during the winter.

Ryan Aeronautical Library

Cabin M-2 with pilot in rear was modified for Dick Robinson whose plane is powered with Ryan-Siemens engine

Westland Oil Company's M-1 was damaged at Terry, Montana, late in 1927 when a pilot taxied his OX-5 Eaglerock biplane into it, chewing off a wing at the fuselage.

During the winter, Westland's pilot/mechanic, Cecil Shupe, rebuilt the M-1, making it into a cabin job and at the same time extending the wing from 36 feet to 40 feet. In those days pilots were inclined to customize airplanes to their own ideas without much formal approval.

When Dick Robinson gave up his partially completed *Gold Bug* B-1 to Frank Hawks he wanted to order another Brougham, but while waiting for it got hold of an M-2 for which Claude Ryan sold him a 125 h.p. Ryan Siemens engine. Some said the airframe was from the old *Bluebird*. In any case the M-2 was rebuilt into an enclosed cabin plane by Robinson's pilot, H. J. (Jack) Laas. The pilot flew from the rear seat as in other M-2s, not from the front as in the *Bluebird*. Robinson had *Hotel El Cortez* painted on the side and the plane became known as the *El Cortez Special*.

Cecil Shupe

Hisso M-1 was converted to M-2 cabin plane by Cecil Shupe, right

Another customized, partially enclosed M-1 type was the M-3C *Spirit of Hollywood,* also known as the *Spirit of Universal Pictures*. A special windshield was built to enclose the space between the top of the fuselage and the bottom of the wing. A glass enclosure also extended around the sides of the front cockpit between fuselage and wing. The pilot flew from the open rear cockpit.

The M-3C, which also appears to have had an extended wing, was later registered under license 10026. Reportedly it was once owned by Howard Hughes and flown in motion picture work by movie pilot Frank Tomick. It replaced another M-1/M-2 camera plane, c/n 18, licensed as 3393, which had been involved in a crack-up during the final filming of "Hell's Angels" late in 1928.

(Early Ryan monoplanes which had the letter "C" after M-1 possibly designated enclosed cabin versions. Thus the Westland Oil M-1 which was rebuilt had a nameplate engraved "Serial No. 23, Type M-1C." The rudder insignia meanwhile had been changed from M-1 to M-2.)

"El Cortez" cabin M-2 is parked just beyond Maddux Air Lines' Ford trimotor

Title Insurance & Trust Co.

Title Insurance & Trust Co.

New name appears on Ryan hangar as B. F. Mahoney takes over

Mahoney Takes Over

ALTHOUGH Charlie Lindbergh and Franklin Mahoney had become good friends during the time the *Spirit* was being built, Lindbergh – in Paris and under tremendous pressure from the adulation of kings and commoners – was less than enthusiastic about having Mahoney join him abroad.

But Mahoney, pushed somewhat by A. J. Edwards, was anxious to bask in the reflected glory of Slim's superb accomplishment.

On the day of Lindbergh's arrival in Paris publicist Richard R. (Dick) Blythe, of Bruno and Blythe, and Mahoney joined in cabling Slim at the American Embassy:

> GREAT STUFF. CONGRATULATIONS. DO NOT SIGN ANY CONTRACTS IN PARIS. EVERYTHING HERE GOING FINE.

Three days later Mahoney cabled again:

> OF UPMOST IMPORTANCE THAT I JOIN YOU IN PARIS. SITUATION IN NEW YORK AND ST. LOUIS MAKES IT ADVISABLE THAT I CONFER WITH YOU BEFORE YOU SIGN ANY CONTRACTS OR AUTHORIZE ANYONE TO ACT FOR YOU. UPON RECEIPT OF CABLE FROM YOU WILL SAIL IMMEDIATELY FOR PARIS. BY HOLDING EVERYTHING IN ABEYANCE YOU WILL AVOID POSSIBLE DISTASTEFUL SITUATION HERE. AS YOU KNOW MY INTEREST YOUR WELFARE PARAMOUNT. CABLE IMMEDIATELY.

Back came the reply:

> MAHONEY
> GARDEN CITY
> LONG ISLAND
>
> AM INSPECTING EUROPEAN AERONAUTICAL INDUSTRY BEFORE REACHING AMERICA THEREFORE USELESS COME OVER
>
> LINDBERGH

Mahoney, apparently could not take a hint. The next thing Lindbergh knew, Mahoney was on his way:

LINDBERGH
CARE AMBASSADOR HERRICK
PARIS

FRANKLIN LEFT MAURITANIA TODAY TO KEEP YOU ENTERTAINED.

HARTSON WRIGHT AERO

Meantime Lindbergh had cabled Mahoney again to advise him that "KNIGHT (one of his backers from St. Louis) IN CHARGE AMERICAN ARRANGEMENTS", but Mahoney was already aboard the S. S. Mauretania.

Tom Mathews, the Ryan publicist, had some insight to what was going on. "When Edwards urged Mahoney to follow Lindbergh to Paris," he wrote, "in order to further Mahoney's exposure, Harry A. Bruno (of Bruno and Blythe) was against it; Lindbergh was against it. But Mahoney got Hartson of Wright to help him and off he went, arriving about the time President Coolidge had the Cruiser Memphis waiting to bring Lindbergh and the *Spirit* home.

"What to do about Mahoney? Through Navy friends and, I think, Lindbergh's good word, he came back on the boat with Lindbergh.

"It did identify the Mahoney name to some extent with the flight, the plane and Lindbergh, but in retrospect I doubt that it was all that important.

"The 'distasteful situation' to which Mahoney had referred in his cable was the thousand and one hysterical propositions for Lindbergh being pushed through every conceivable channel, each offer more fabulous than the last, with plenty of fall-out for anybody closing the deal."

It was three weeks after his arrival in Paris before Lindbergh set foot back in the United States and then the enthusiastic receptions started all over again.

It was late June before Mahoney returned to San Diego. He had been gone eight weeks.

At a civic dinner honoring Mahoney, "Chamber of Commerce executives," as reported in the press, "predicted a brilliant future for the Ryan Airlines, comparing it to the Douglas factory in Santa Monica which grew in the space of two years from a plant employing 150 men to one of 750 men following the success of the Round the World fliers in Douglas planes."

But a valuable eight weeks had been lost while the Ryan company foundered with no one at the helm.

From the day of Lindbergh's landing in Paris inquiries about Ryan airplanes came in by the hundreds — by phone, letter and by wire. The company was just not equipped to face the deluge.

A. J. Edwards had been taken to the hospital in Los Angeles with a fractured skull after Doug Kelley tore up the *Bluebird* on landing. Hawley Bowlus wanted to see Edwards and get some direction on what they should be doing about all the inquiries. Edwards suggested Bowlus fly up and bring along his and Mahoney's secretary, Georgia Mathias.

"Hawley wasn't the world's best pilot," Georgia recalls, "and the flowers I took along from Mahoney's mother for Edwards might better have been for me. It was a wild flight; Hawley flew just above the waves along the coast where it parallels the highway. The wheels were wet when we landed in Los Angeles.

"When we got to the hospital, photographers from the newspapers were there and took pictures of A. J. with his bed just covered with telegrams. After taking dictation for several hours I flew back

Bunnell Photo

A. J. Edwards ended up with a skull fracture

H. A. Erickson

Figuratively as well as literally Mahoney is in the driver's seat as he poses with his Stutz Bearcat in front of a Hisso-powered Brougham

H. A. Erickson

Moving spirits behind the "Spirit" were Bowlus, Mahoney, Lindbergh and Hall, left to right

Bunnell Photo

Mrs. Jennie L. Mahoney and friend with a Hisso-powered Brougham

with Hawley to San Diego. We had a forced landing en route and a rougher landing on arrival at Ryan Field on Dutch Flats."

Not only was Edwards temporarily out of circulation but Claude Ryan had left Ryan Airlines, Inc. just before Lindbergh's flight to go into the business of importing German aircraft engines under the Ryan-Siemens name.

Who was in charge of the Ryan aircraft business? Until Mahoney's return Edwards struggled with the business and sales problems; Bowlus with production as best they could.

Once back in San Diego, Mahoney made up for lost time. His attorney, William G. Mirow, was already working on plans for the formation of a new corporation. It was to be known as Ryan Airlines, Inc. for, in fact, Mahoney earlier had never found the time, while he was a partner of Claude Ryan's, to sign papers necessary to properly incorporate the firm.

Mirow's plan of June 30 called for a capitalization of $500,000 with 2500 preferred shares and 2500 common shares, each of $100 par value. After issuance of preferred shares to Franklin Mahoney equal in par value to the tangible assets of approximately $50,000, he would receive one share of common stock for each preferred share sold to the public.

Within two weeks, however, the initial plan was abandoned and instead articles of incorporation of the reorganized company, under the name B. F. Mahoney Aircraft Corporation, were filed July 15, 1927. The company, capitalized at $500,000, announced it would "continue the manufacture of monoplanes, to be known hence forth as Mahoney monoplanes, instead of the former name of Ryan monoplanes."

The board of directors held its first meeting on July 25. It consisted of B. F. Mahoney, his mother, Mrs. Jennie L. Mahoney, Richard T. Robinson, Jr., W. H. Bowlus, factory superintendent, and A. J. Edwards, director of sales.

Prospects could hardly be brighter. For every order that the company was accepting, Mahoney reported, they were forced to turn down five others. "We could be selling five planes every day now," Mahoney said, "if we could only build them, so great is the demand for commercial craft at the present."

A chain of sales organizations was to be established by Sales Director Edwards in practically every nation of Europe, Latin America and Asia as well as the United States. Before the end of 1927 the Mahoney plant expected to double its personnel and was making active preparations for enlargement of the factory.

* * *

Brougham production began to swing into some semblance of order once the Mahoney factory at the old fish cannery standardized on the Wright Whirlwind engine installation. Fuselages were towed behind an open roadster out to the Dutch Flats field. Wings built at the cannery were carried on a flatbed trailer, arriving at the hangar at 9 a.m. to be mated to the fuselage. The engine would be installed, the ship rigged and the plane test flown at 4 p.m. the same day.

The first Whirlwind-powered Brougham to be completed after Frank Hawks' prototype B-1 was Serial No. 32 (c/n 32)

Donald A. Hall

Wing for "Spirit of St. Louis" is moved from fish cannery factory for transport to Dutch Flats flying field

Bill Monday

"Queen of the Yukon" at Ely, Nevada. Clyde C. Wann, in golf knickers. To his left, pilot Bill Monday

with license 3113. It had been ordered July 8 by Clyde C. Wann of Portland, Oregon, vice president of the Yukon Airways and Exploration Co., Ltd., who some ten years earlier had established himself in business at Whitehorse on the Yukon.

Pilot and part owner of the new Brougham, *Queen of the Yukon,* was A. D. Cruickshank, district representative of the Northwest Mounted Police. The plane was to be used between Whitehorse and Dawson to carry mail, freight and passengers over a 550 mile route requiring from eleven to fourteen days by dog sled and river steamer. Because of the many lakes, the plan was to equip the plane with pontoons for water landings and skis for snow, as well as the normal landing gear.

With the new interest in flying which the *Spirit of St. Louis* had generated, Cruickshank decided to barnstorm the Northwest carrying passengers in his new $10,000 Brougham—the first cabin plane used in that area on local sightseeing flights — before heading for Alaska. The only problem was that Cruickshank was not a very good or experienced pilot, having considerable difficulty handling the Brougham on the ground.

After a landing at Ely, Nevada, in which the *Queen of the Yukon* got mired in the mud, Cruickshank asked for help. The Ryan plant got hold of one of its old pilots, Bill Monday, and sent him off under contract to fly for Cruickshank.

For ten weeks they barnstormed through Nevada, Northern California, Oregon and Washington, carrying 1600 passengers on 464 flights and clearing some $8000. At Reno, Monday reported he landed in eight inches of mud but later was hauling full loads from the same field, 7000 feet above sea level. Monday left them at Vancouver, British Columbia. Because it was so late in the year, the *Queen* was put aboard ship for the trip to Skagway, Alaska.

On November 11, 1927 service was begun from Whitehorse to Dawson, capital of the Yukon Territory, with stops at Mayo and Keno. Occasional flights were also made to Skagway. Although having no contract with the Canadian Post Office, the airline charged 25 cents on its own account, and reportedly issued its own airmail stamps, for each letter carried.

Yukon Airways also had the distinction of being able to claim that the *Queen* was the only commercial plane equipped with charcoal foot warmers. Photos in an old Mahoney brochure of the plane on skis, carrying Canadian license G-CAHR, describe it as "one of the Ryans used in the Yukon gold mining region where the ship is operated at temperatures ranging from 30 to 50 degrees below zero."

According to R.E.G. Davies, author of "A History of the World's Airlines," the Ryan Brougham of Yukon Airways was wrecked at Mayo on April 5, 1929.

Another very early Brougham—it still used rubber "bungee" shocks — was c/n 39, the *Pride of Pittsburgh.* Under licence 3257 it was the first of a small fleet of Broughams bought through Mayer Aircraft which Clifford Ball, Inc. operated on Contract Air Mail route No. 11 between Cleveland and Pittsburgh.

Robertson Aircraft Corp., operator of C.A.M. 2, the St. Louis-Chicago airmail route which had been flown by then

"Pride of Pittsburgh" was purchased for Cleveland airmail route. Ryan publicist Tom Mathews in dark suit

Clifford Ball

H. A. (Bud) Gurney, fellow pilot of Lindbergh's on St. Louis-Chicago mail run, left, with newspaperman Roy Alexander who later became TIME managing editor

H. A. Gurney

Capt. Charles A. Lindbergh, purchased Brougham C-958 (c/n 37) for passenger service between the two cities. It was decided to get the plane after the company reported it had recently "broken all records by carrying a passenger on every trip over a period of a week."

Later, additional Broughams were added to the Robertson fleet at a time when charter flights accounted for almost as much flying time as the scheduled operations. Harlan A. (Bud) Gurney, one of the Robertson pilots and a flying buddy of Lindbergh, explained:

"Ryan Brougham C-958 was sometimes used on the St. Louis to Chicago CAM #2 Air Mail, but mostly it was flown on charters. In those years before wirephoto, pictures and movie news films were flown to Chicago and to New York quite regularly.

"I flew most of those trips and mostly at night when few lighted fields existed. There was no radio either so navigation was a matter of well planned dead reckoning plus a good magnetic compass. Parachute flares were used to land at unlighted fields.

"A pile of newspapers weighing 700 pounds was flown daily from St. Louis to Houston, Texas each of four nights during the Democratic Convention that nominated Al Smith for President. I landed only at Little Rock, Arkansas for fuel and returned from Houston with Pathe news reels and Acme Photos. It was seven hours each way so a round trip a day slightly fractured present pilot duty hour regulations.

"You may notice that the tail surfaces on 958 were different from the original.

Robertson Aircraft operated eighth Brougham built on CAM No. 2. Pilots are Leslie Smith and Bud Gurney, right

H. A. Gurney

Originally the ship was too unstable to enable extended instrument flight — as was the *Spirit of St. Louis.* We extended the engine mounting and enlarged the fin and stablizers to make the aircraft a much better one.

"I don't know what eventually became of old C-958 but it was one of the very first Ryan Broughams with a Wright J-5 engine. I had quite a few real rough rides in it and one that I'll never forget. It was in a tornado condition about 2:00 a.m. that night in 1927 when Advance, Missouri was blown apart. I found there was weather I couldn't fly. It flew me!

"Out of all control, I first saw lights through the top of the front windshield; leveled out by lightning flashes and spotted a field where we had barnstormed a couple of years before. Then used a flare and landed in a flooded field ok. Any other ship would have come apart in the violent winds that night. Old 958 stayed in one piece while the seat belt pulled such negative Gs that I had raw welts across my lap. The Ryan Broughams could really take it! Nothing else would have!

"Robertson would book flights to anywhere at any time of day or night; a doctor to some emergency in northern Iowa. Fly there, find an open field with the first flare, pull the second one and land toward its light after a tight 180° turn. It always worked! (I'm here.)

"Once old 958 picked up about four inches of ice but without the problems we had with just traces on a deHaviland or the J-5 Fairchild razor-back we also flew. The Travel Air mail ships were even worse; the aileron struts would vibrate and break loose. Old 958 was a good ship!"

* * *

The B. F. Mahoney Aircraft Corporation had filed articles of incorporation July 15, 1927 but it was October 6 before approval could be obtained from the State Corporation Commission to sell 5,000 shares of its $100 par value stock. Even then Mahoney was required to place his stock in escrow.

A July 31 financial statement showed the company was insolvent. It had a negative net worth of $9,865 since current liabilities of $84,051 were in excess of the $74,186 in current assets. "The business," the statement said, "also has certain intangible assets of the alleged value of $285,510." Included in this figure were $510 in prepaid insurance and $285,000 in good-will. The value of airplane designs was computed at $25,000 for the M-2, $100,000 for the B-1 and $150,000 for the NYP. Trade-names were valued at $3,000 each for "Ryan Airlines" and "Ryan Flying Co." and $4,000 for "Los Angeles-San Diego Air Line." In exchange for the intangible assets — the reputation of its planes — B. F. Mahoney was to receive 2756 shares valued at $275,600 ($285,510 in intangible assets less the negative net worth of $9,865).

Investors would be offered the remaining 2244 shares, of the 5000 shares authorized, at $100 per share,. Obviously, the company sorely needed the two hundred thousand dollars which would be raised from sale of stock for expansion and working capital.

Customers were standing at the door waiting to buy airplanes but the company simply was not geared to handle the volume of available orders on a business-like basis. Nor were the engineering problems being solved. Donald Hall had worn himself out under the stress of the 60-day deadline to build Lindbergh's plane.

Before joining Ryan Airlines in January 1927 Hall had worked for Art Mankey at Douglas. "He was," said Art, "a wonderful man, a wonderful engineer and a wonderful stress analyst, but he was the slowest man I ever saw in stress analysis work.

"After Jack Northrop and I had finished work on the M-1 for Claude Ryan, I suggested to Don Hall that he ought to go down to San Diego as I knew they needed a fulltime engineer. That's how Don arrived on the scene just before Lindbergh came along looking for a plane for the trans-Atlantic flight. But he deserves every bit of the credit he got for engineering work on the *Spirit of St. Louis.*

Department of Commerce took delivery of c/n 48, first Brougham licensed with full technical data

H. A. Erickson

"After the flight I don't know whether Don got tempermental or was just completely exhausted — at least he was worn out emotionally. So he went to a camp way up in the mountains. I guess the factory was in touch with him, but he couldn't contribute much from his mountain hideaway just at the time the rush of orders was flooding in.

"One of the orders — a very important one — was from the Department of Commerce. The plane was to be Serial No. 48 and it would be used by Clarence M. Young, head of the Aeronautics Branch. New regulations required full engineering data on licensed aircraft, but no one at the Mahoney plant had run a complete stress analysis or other technical data on the B-1 Brougham.

"The plant was in deep trouble. Don still hadn't returned to work. It was then that Franklin Mahoney got hold of me at Douglas and asked me to come down to San Diego as chief engineer.

"I was making $75 a week at Douglas. Franklin offered me $85 to start and $100 a week after three months. We put the crew to work and whipped out the technical data so that everything was approved by the time Clarence Young took delivery in October. No. 48 became NS-15 and its was the first Brougham licensed with full technical data. (Later the B-1 model was granted Approved Type Certificate No. 25, January 1928).

"When I came down as chief engineer I was disturbed about Don and asked Mahoney, 'What the hell are you going to do about Don?' He assured me he'd take care of everything.

"One day I was out at the Dutch Flats field when Hall showed up. He'd just returned. Innocently he asked me how long I was going to be in San Diego.

"I said, 'Don, have you seen Franklin Mahoney yet?'

" 'Yes,' Don responded, "I had quite a talk with him.'

"Did he tell you that I'm now chief engineer?

" 'No, he didn't tell me.'

"Then you better go back and ask him."

"The upshot was that we made Don chief experimental engineer."

Hall, according to his assistant, Wilhelm F. (Bill) Schult, didn't want to have anything to do with the Brougham because no adequate detailed drawings had been completed and because planes were being built so fast they couldn't use one of the airplanes long enough to calculate the center of gravity for their engineering analysis work. Too, Hall was disturbed about Mahoney's need to sell airplanes in order to use the cash to continue to get another engine out of bonded storage for each plane as it moved along the production line.

Lacking the type of production control which later became standard practice in the aerospace industry, Ryan craftsmen modified design features of the B-1 as production continued.

The *Gold Bug* for Frank Hawks was the prototype of the B-1 series. Its wing-to-strut-attachments were covered with a streamlined fairing, but subsequent planes did not have this feature, nor were the landing lights in the wing leading edge continued. In due time the tail skid was replaced by a tail wheel in later B-1s. The large fairing over the rubber bungee landing gear shocks gave way to the much smaller streamlined air-oil shock struts.

The early five-segment windshield was improved by cutting two below-centerline sectional windows for improved downward vision. Early models had the door only on the right side. A door was later added on the left side of the plane to provide access for rear seat passengers.

Instead of getting busy on engineering problems with the Brougham, Hall went to work on a new concept for a two-place low-wing plane of unique design. It would be the first to have a "flying tail." In place of conventional fixed horizontal and vertical tail surfaces forward of the elevators and fin, the entire tail surfaces of the X-1 experimental job would be moveable — a "slab" tail.

But more later about the Mahoney-Ryan Special which Hall thought could play a major role in the company's future. Its fate, however, was to be dubbed the *Doodle Bug* and become known as *Hall's Folly*.

Hall's new concept . . . a monoplane with flying tail . . . irreverently called Doodle Bug

H. A. Erickson

H. A. Erickson

Pot-bellied Brougham with 46-foot wing was built for projected flight to Rome

The Special Jobs

THE history of Brougham Serial No. 36 remains somewhat of a mystery. Many people have vague recollections of it, but no one seems to have the full story. It was a "Japanese copy," an exact duplicate, of Lindbergh's plane – for the Japanese. Like the *Spirit* it had the large 46-foot wing with ailerons inboard of the wing tips.

Strangely enough, not a word about the plane seems to have appeared in the aviation trade publications of that period.

Ed Morrow states that "the NYP-2, as it was called, was all-silver. Red Harrigan flew it but did not like its handling characteristics. The Japanese had been in San Diego to watch its construction. It was crated and shipped to Japan in late summer 1927."

Photographs of NYP-2 taken in Japan the following year show it with license J-BACC. On the rudder was inscribed "Tonichi Daimai No. 11."

In June 1970, Haruo Takaishi, General Manager of the Aviation Department of the Mainichi Newspapers, filled in much of the background of the Japanese duplicate for us.

"The inscription signified that this was the eleventh airplane to be purchased jointly by Tonichi and Damai, abbreviations for our newspapers in Tokyo and Osaka," he wrote.

"On April 28, 1928, J-BACC, piloted by Fumio Habuto, established a new Japanese distance and endurance record by flying 2000 kilometers in a 13 hour, 23 minute dawn-to-dusk non-stop flight on a circular course over central Japan. Because of rather poor weather, Habuto flew most of the time at an altitude of little more than a thousand feet."

After the flight, Habuto was hailed as the Japanese Lindbergh, but little is known of the further use of the plane by the newspaper sponsoring the record flight. "In 1938," Takaishi writes, "J-BACC was scrapped because of the limited capacity of the hangar."

Perhaps the most intriguing story about the duplicate of the *Spirit* is that the Japanese in turn reportedly duplicated J-BACC – double size – for a trans-Pacific flight proposed in August 1927 by the Teikoku Aviation Association which raised the funds by popular subscription.

If true, this would have been the K-12 monoplane whose manufacture was

NYP-2 was a duplicate of Lindbergh's plane for Japanese newspaper

ライアン NYP-2 型長距離機　リンドバーグが大西洋横断に成功した 305
機体と同型機を昭和2年毎日新聞社が購入した。発動機はライト・ホワールウィンド J-5C 空冷式星型 200 hp をつけ，胴体中央に大主燃料槽をつけて，

Mainichi Newspapers

Scaled-up Japanese long-range Brougham-type monoplane had 62½-foot wing span

Mainichi Newspapers

Kawanishi K-12 monoplane flies past pagoda on Japanese coastline

Mainichi Newspapers

started in October 1927 and completed in June 1928 by the Kawanishi Aircraft Works of Osaka.

With a wing span of 62½ feet compared with the 46 feet of J-BACC, the Kawanishi K-12 required an engine of double the horsepower — a 500 h.p. BMW (Bavarian Motor Works) water-cooled engine. Test flights were terminated when the government aviation agency grounded it for "insufficient power and flying range.' Plans for the trans-Pacific flight had to be abandoned.

A second K-12 had been built and the pilot reportedly killed when the overloaded plane was destroyed in a fiery crash on takeoff.

Tom Mathews has this recollection of the "Rucky Rindy" duplicates:

"I think we shipped out two airplanes to Japan after the Lindbergh flight, a regular Brougham and the duplicate of the *Spirit*. One of the boys from the factory went along to supervise setting the ships up and rigging them. About six months after he returned a pilot came through San Diego from Japan and said the Japanese were building a duplicate — even duplicating a dent in the cowling that had occurred during shipping. He though the ship was much bigger than the one we had shipped out.

"As you can see from the picture which appeared in Aviation Magazine in June 1928 the cockpit is in front and a water radiator sticks out the side to further complicate the airflow over the tail group, which was on the too-small side, anyway."

H. A. Erickson

Special Brougham to carry M-G-M Lion takes off from San Diego bound for New York

Along with production of standard B-1 Broughams for the expanding flying service and business-owner market, the fledgling Mahoney organization had to weave its way through a continuing series of requests for 'special' models for a variety of publicity flights.

Someone came up with the bright idea that Leo, the famed MGM lion which appeared on the screen at the opening of all Metro-Goldwyn-Mayer films, should make a non-stop flight, too.

In August, Louis B. Mayer, vice president of MGM, entered into an agreement, backed by a $5,000 payment, with B. F. Mahoney of Ryan Airlines to charter a special Brougham for the publicity flight. Its legalese wording was good for some laughs:

"You agree to accept from us one live lion, and you agree to transport the same to Roosevelt Field, New York, by non-stop airplane flight," and later . . .

"You agree that the cage in which said lion is to be kept while on said airplane will be so placed and constructed that at least the head of said lion will be clearly visible from the outside of the plane and between the struts of said plane.

"The official start of said non-stop flight is to occur not earlier than September 8, 1927, nor later than September 15, 1927."

But who would undertake a thirty-hour flight with a hungry lion for a passenger? A. J. Edwards had an idea that Martin Jensen might be available.

Jensen, an early San Diego pilot and friend of Claude Ryan's, had gone to Hawaii early in 1927 to fly a Ryan Standard biplane Claude had sold to the owner of a sightseeing firm there. After Lindbergh's flight the Dole Pineapple Company offered $35,000 in prizes for the first non-stop flight from the West Coast to Hawaii. Jensen came back from Hawaii and obtained a monoplane, similar to the Brougham, made by Vance Breese. His 'Aloha' was the second plane in the Dole Race to land in Hawaii after leaving Oakland, California August 16, 1927. Art Goebel in the *Woolaroc* was there ahead of him.

"Several days later," recalls Jensen, "a cablegram from A. J. Edwards of Mahoney Aircraft reached me in Honolulu offering transportation to San Diego and the chance to fly a special Brougham non-stop from Southern California to New York, carrying with me as a passenger MGM's trademark lion. After delivering Leo I was to enter the plane in the National Air Derby non-stop race from New York to Spokane, Washington, and later fly the lion to theaters around the country.

"When I arrived in San Diego the plane, which Mahoney was leasing to MGM, was near completion. There was an iron cage about four by eight feet in the fuselage behind the pilot with plate glass sides so Leo could pose for his picture while in flight. The wing had the standard 42-foot span. Extra fuel and oil for the long flight were carried in a bulge under the cage."

The agreement between Mahoney and Jensen worked out September 15 provided Jensen a salary of $100 a week for five weeks and one-half of the prize money if the Brougham, entered in the factory's name, should be among the winners in the race to Spokane.

Take-off was to be made for New York no later than September 15 and from there for Spokane no later than September 24. (Two who signed the agreement as witnesses were newspaperman Max Miller, and Fred H. Rohr, who was to head his own aircraft firm some 15 years later).

M.G.M.

Bars of Leo-the-Lion's cage aboard Brougham are seen in this photo of Martin Jensen, pilot and B. F. Mahoney

"I took off from San Diego at ten o'clock on Friday morning September 16, "Jensen continued, "with 400 gallons of gasoline, from the same field on Camp Kearny where Lindbergh had run the fuel tests four months earlier on the *Spirit of St. Louis.*

"The takeoff of the special Brougham, c/n 42, license 1550, was uneventful except for center-of-gravity changes due to my passenger who was not wearing a seat belt, but Leo soon settled down.

"I had difficulty getting altitude to cross the Laguna Mountains east of San Diego, then in the Imperial Valley where the temperature was extremely high, the plane settled to about 100 feet above ground — not a very comfortable feeling.

"Northeast of Phoenix I was back up to 3,500 feet but having trouble getting altitude. At Gisela, Arizona, I was probing for a place to get through the mountains up Tonto Creek and over the Mogollon Rim but soon found myself in a canyon losing altitude.

"There was a slit in the mountain but it was probably narrower than the wing span of the plane so I elected to turn and go up a blind canyon full of oak trees. I hit them going wide open — at least 125 miles an hour. The flares on the landing gear ignited and went off, the gas tank broke open, the wings were torn off, the engine broke away. The last tree we hit was at least 10 inches in diameter and we ripped it right out of the ground. But, miraculously, neither Leo nor I were seriously hurt since much of the fuselage was intact.

"All the windows were broken and I climbed out through the windshield. I made a beeline for the first big tree I could find and climbed it as I wasn't too sure how Leo had made out in the crash. After a while he quit growling in his cage and I went back for my gun, but when he started snarling again I retreated to my position in the tree.

"Well, I couldn't stay there forever so after a while I mustered up enough courage to get back in the airplane and try and recover the compass. Then I climbed to the top of a nearby hill to survey the area for the best way out, knowing I should go downstream and would eventually find some kind of habitation in the direction of Gisela.

"I started down Hell's Canyon — later to be known as Leo Canyon — which ran into Haigler Creek and it, in turn, ran into Tonto Creek. Scrambling down steep rocks and vertical canyons, often hanging on almost by my fingernails, darkness soon overtook me and I knew I shouldn't continue in such treacherous terrain much longer.

"I lay down in a small gully, tying the canteen straps around my feet so I wouldn't roll off during the night. My misery doubled when it started to rain. Fortunately I had brought along strips of the plane fabric which had been torn off in the trees. I took off my leather jacket, shirt and undershirt and wound the plane fabric tightly around my torso before putting my clothing back on. What a miserable, cold, wet night!

"At daybreak I was ready to move on. I got up, walked a few feet and there was the creek below me — a hundred feet, almost straight down!

"Following Haigler Creek I had to get into the water up to my armpits — it was ice cold and I hurt all over. Just

Martin Jensen, second-place winner of Dole air race to Hawaii, was hired as pilot of Brougham to publicize M.G.M.'s lion.

Martin Jensen

Martin Jensen

... the wings were torn off, the engine broke away, the flares ignited ...

then I heard a heck of a racket in back of me. Leo? When I turned around it wasn't Leo but a wall of water from the unseasonal rain! I took off up a steep shale hill to get out of there; two-thirds of the way up I lost my footing and slid down to the bottom. A second try over the shale was no better. Finally I made it over some big rocks to Tonto Creek but not until having encountered a dozen rattlesnakes on top of one boulder.

"At last, from the top of a hill I saw some cattle. Being raised on a farm I knew that if they were startled or chased they'd go home, and I could follow them to civilization. Only after I'd chased them clear around the mountain did it dawn on me that they were range cattle; they didn't have a home.

"Guided since late Friday by the sun, moon and stars I was headed west again off the mountain when I ran into a pack of coyotes right in my path. I decided I'd scare them off when I got within a hundred yards of them, but a shot from my gun disturbed them not at all. I wasn't about to argue with them so climbed the one tree that was there, thinking they'd soon leave and I could resume my trek by moonlight.

"I stayed up in the top of that tree all Saturday night, laying over a branch, hanging this way and that trying to get at least some sleep. At dawn the coyotes were still there. Then along came a duck, quacking as he waddled off in the direction of the coyotes. That convinced me that if a duck could get away with it I could.

"Through gullies, up hills, through water and over rocks. That's how I spent Sunday, the third day without food, in the open. But fortunately I had filled my canteen everytime I left the creek. By evening I was on top another mountain. I made a bed out of boughs from a fir tree and lay down for my first real sleep in three nights, with my gun strapped to my hand in case I got into a scuffle with a wild animal.

"Monday morning I could barely walk. I got down on my hands and knees and crawled along until I came to a cow path. There I stood up and then worked my way to the top of a hill. I smelled smoke and saw a farm in the distance. As soon as I thought someone might hear I began shouting. Finally two fellows spotted me. One of them carried me piggy-back down the hill. They cut the boots off my swollen feet, bandaged them with a bedsheet, gave me some food and put me on a burro to get me to the nearest automobile.

"The Model T Ford had a flat tire which I helped fix; then water had to be obtained for the leaky radiator. We finally made it to the nearest Ranger Station, where we were informed we couldn't use the phone because 'somebody's lost out there, and we have to keep the line open.'

"When I told them who I was I finally got to use the phone to San Diego.

" 'This is Jensen,' " I said.

" 'Yes; how's the lion?' " they said.

"When I told them about the situation in which I left Leo they explained that MGM had said to spare no expense in getting Leo out. Under the circumstances I found it a little tough playing second fiddle in their concern for Leo.

"There was a hunter at the Ranger Station who seemed extremely interested in the lion. He was an attorney and explained there was no need for me to go back for Leo since no one was with him and he was therefore claiming the lion on the basis of salvage. After all, he said, I'd abandoned Leo.

"I told him I was coming just the same, but then he reminded me that he'd

have a gun. "Well, I replied, 'what do you think we are going to carry, billy clubs!'

"Two days later I led a posse of farmers back in on horseback from Payson. Our attorney friend was already there. I told him he could go on now that we were taking over. He got up and left. Just that easy.

"Leo was in bad shape. It had been seven days since he had been fed and watered. The cage was on a 45 degree angle so the lion couldn't stand. He'd been cut up with flying glass and the flies and maggots were driving him crazy. Sam Haugh, the farmer who headed the posse, swabbed him down with sheep dip. Leo seemed to like that.

"We worked the cage over to a flat area and set up camp for the night. Leo was fed the calf which had been slaughtered on the way in. He was a happier lion but whenever he roared during the night the whole canyon reverberated.

"We made a sled out of the fork of a tree, lashed the cage onto it with chains and put two burley mules to work pulling it out of there. Five men had to support the cage as it was pulled at an angle along the slope of a hill. Then one man stepped on the lion's tail. Leo let out a roar. The mules took off. The cage dumped over onto one of the men and started rolling.

"Finally we got down to where an old Dodge truck was parked. There the cage was loaded onto the flat bed and trucked into Payson where Leo was displayed in the garage for the benefit of children in the area who had been dismissed from school.

"The postmaster finally took word of the crash and rescue to Globe, the county seat, and from there the news was finally phoned to Phoenix and the outside world.

"By now Leo was weak and docile. In a few days a circus truck came for him, and eventually he emigrated to Australia where he was displayed."

Martin Jensen

. . . the cage was trucked into Payson where Leo was displayed . . .

. . . two burley mules pulled the cage out of there on a home-made sled. . .

Martin Jensen

Several days later Jensen was in Tucson where he shared the spotlight at a civic affair with an even more famous pilot of Ryan Broughams — Charles A. Lindbergh — who was then touring the country to promote aviation.

Recalling the crash many years later, Jensen insisted that "it was not due to either airplane or engine failure but was due to my error alone."

Jensen was too modest. His error occurred not at the scene of the accident but in San Diego. He probably should never have taken off at all. The wing loading was very high and Ed Morrow reports that "the design of the propeller forced us to install a broader front engine cowl than was proper for adequate cooling in the hot air conditions then existing over the desert.

"Engine overheating was Martin's problem. His minimum altitude over the desert should have been well above 10,000 feet which should have been attained while crossing over the mountains and before reaching the flat, hot country." Unfortunately the plane just didn't have the necessary 'stuff' for the flight.

The publicity venture was less than a great success for both MGM and Mahoney. Not only had the Mahoney company lost its airplane but had to refund MGM about $2,200 since the agreement provided that "you are to refund to us an amount equal to $1.00 per mile for each mile between the point where the plane has actually landed and Roosevelt Field, New York."

In time Leo no doubt passed on to lion heaven, but 45 years after the special Brougham crashed, Martin Jensen was still living happily in San Diego.

* * *

The most entertaining, if perhaps somewhat fictionalized, account of the saga of Martin and Leo came from that ebullient raconteur publicist, Tom Mathews —

Martin's contract to pilot MGM's Leo "The Flying Lion" was three weeks of calculated hysteria. Both Leo the Flying Lion and Martin somehow lived through it. Conceived by Howard Deitz, MGM's brilliant "idea" boy, it was to be a "follow-on" to the Lindbergh transcontinental flight covering the same route, in the same type plane, this time with Leo, the Flying Lion, and Martin taking Lindbergh's place at the controls.

The West Coast activity was turned over to MGM's Joe Sherman who thought the first order of business (after ordering the plane) was to bring Leo down to San Diego and introduce him to his pilot. If everything went well, his thinking was that Leo would move in with Martin, and Marguerite (Mrs. Jensen) could move into a hotel while the plane was getting finished. When Marguerite said no to that, Martin looked relieved.

Because of Leo's size and the lack of space in the airplane, his open bar cage was placed right up against the pilot's seat in this converted Brougham. On the first test flight Leo faced forward. He was lying down until the ship got airborne, then he stood up and Martin said he began licking the back of his neck. He thought the lion had gotten loose but quickly saw it was just his nose sticking out through the bars. Then Leo let out a roar and Martin quickly landed and suggested the lion would be more comfortable facing the rear. So they turned Leo around. He climbed in and out of the airplane like an old time pilot. (Leo, not Martin. Martin took his time).

On the second flight everything went well until they hit a little rough air and Leo stood up again and stuck his tail out through the bars, wrapping it around Martin's neck and waving it around in front of his face. Martin landed again and suggested a board of some kind in the end of the cage to keep Leo out of the wind as Martin sometimes flew with the side vent open. This was done and the big day arrived.

The take-off was from Kearny Mesa, a 5,000-foot dirt strip used by the Navy and Marines as a parade ground and auxiliary field. Four hundred fifty pounds of lion, 450 gallons of gas and Martin Jensen. They used the whole 5,000 feet and mushed out over the low hills to the east, never to be seen again, or at least that is what we were beginning to fear after four days.

There were plenty of rumors, mostly about Leo the Lion seen roaming around Ohio, Mexico, Montana, Arizona. Marguerite was worried; she knew Martin could put the ship down if he had trouble but had the haunting feeling that he might not have gotten out of the plane before Leo. Finally Joe Sherman took her aside and explained that Leo didn't have any teeth and had been raised from a little baby on milk toast and Pablum. I could only hope that she had not been around at feeding time. His chompers had looked pretty good to me as he tossed off a quarter of beef.

Finally, on Sunday afternoon, Joe Sherman and I went down to Tijuana to have dinner and duck out on the heat from the reporters and news services. In the middle of dinner the flash came in that Martin had walked into a ranch house high in the mountains of Arizona. Joe grabbed the check, tossed some money on the table and started running toward a telephone.

Wilhelm Schult

Hollywood kleig lights, left and right, light up the scene as publicity pictures are posed

"What's the first thing you would do if you had crashed in Arizona with Leo the Flying MGM Lion," he shouted at me.

"I'd jump out and climb a tree," I said.

"No you wouldn't. First you'd make sure Leo wasn't hurt, then you would look for a stream and bring him back some water, feed him your ham sandwiches, and then go for help,"

"I would?'

When Martin could get to a telephone he called Mahoney in San Diego who called Joe Sherman in Tijuana. It was Joe who was briefing the press — the water bit, the sandwich bit, the long struggle down the mountain after making sure Leo was comfortable, a little vague about the exact location to give MGM a chance to get there well ahead of the press and rescue Leo, if Leo was able to be rescued. Leo survived but suffered a bad cold and his roar was almost silent; never did get it back. MGM got a younger Leo.

After everything had quieted down I asked Martin what had happended. It was getting dark and he was just about at his service ceiling with the load he had going through these mountains. He thought he would make it but toward the top the plane started mushing down and when he saw he couldn't make it he cut the switch, shut off the gas and tried to set it into a gulch so the wings would help absorb the shock.

Next thing Martin remembered he was lying about 20 feet from the wreck. He got up, didn't seem to have any broken bones, didn't know if the lion was in or out of the plane, didn't hear any noise.

"What did you do then?"

"I thought I better climb a tree 'til I figured out what to do next."

As a foot note to history it should be recorded that Martin did indeed give Leo all his sandwiches before he left. "I figured I could get out alright," he recalled, "but the lion couldn't."

H. A. Erickson

Carrying an experimental license the pot-bellied Brougham is prepared for a test flight

. . . goggles upside down, Mrs. R. posed with a screwdriver . . .

Tom Mathews \ H. A. Erickson

The sequence of events in the saga of the special Brougham *The Spirit of Holy City Love* is pretty difficult to reconstruct 45 years later.

Old newspaper and magazine articles would give the impression that there may have been two identical pot-bellied Broughams — 'almost duplicates of Lindbergh's NYP' — one for a long-distance flight, the other for an endurance record attempt.

Others, who are more knowledgeable, say there was just the one plane, but their recollection of events does not fit the chronological sequence indicated by the press stories. The latter, however, may well be in error.

This tall tale starts out with 'Father' William E. Riker, founder of Holy City in the Santa Cruz mountains of California, and one of his followers, Mrs. Evelyn Rosencrantz, reportedly a wealthy widow.

Wouldn't a "Holy City to Holy City" flight be a great thing — from California to Washington, D. C., to New York and then non-stop to Rome, the Eternal City? In any case, a special 46-foot wing Brougham was ordered for the long flight.

"The aviatrix," reported Aviation Magazine in its February 20, 1928 issue, "will enter intensive training at San Diego to develop her flying knowledge. The plane is due to be completed on March 20. (Like MGM's 'Leo the

Lion' plane it was one of the first to have the lower quarter windshield glass section, for better pilot visibility downward, which later became standard on all Broughams).

"Before she takes off early in April for the Eternal City, Mrs. Rosencrantz said she would attempt to smash the record flight of 52 hours now held by German airmen." After reaching Washington, according to the article, she would remain in the air until the record was broken.

Tom Mathews, Mahoney publicist, perhaps embellished the story:

"Our usual price of $15,000 for the plane was, in this case, $20,000 for plane and pilot. This was a deal between Father Riker and the factory, but I suspect he was charging the widow more.

"At any rate Mrs. Rosencrantz came down to be near the plane — just like Lindbergh — while it was being completed, and for publicity pictures.

"I put a leather coat on her, together with a leather hat in the style then popular, with the goggles upside down since they fit the hat better that way. She posed in the picture with a screw driver tightening a rocker box on the engine. I could see the headline — Mrs. Rosencrantz to Act as Co-Pilot and Mechanic on Rome Flight.

"For a pilot we found a daredevil from the east coast who was willing to pick up five grand on a little over-the-water gamble.

"Riker paid a deposit and signed the contract, Mrs. Rosencrantz arrived, the pilot arrived and work on the ship was proceeding when we ran into a really big problem.

"It seems Mrs. Rosencrantz wanted the pilot to be sociable while the ship was being made ready, but 'Joe' didn't picture that as part of his job even though our aviatrix though 'we ought to be better acquainted since we'll be hours and hours cooped up in that little cockpit together.'

"No, no! Absolutely not,' said Joe. 'That's not included as part of the five grand to fly her to Rome.'

"Mrs. R. was very distressed, but like all problems it worked itself out. I had the pilot fly a few trips to Los Angeles with her whenever Mrs. R. felt so inclined, and otherwise kept them apart. What happened on the way to Rome was up to them.

"Riker had paid ten grand down but was slow coming up with the second ten grand. He explained that Mrs. R. hadn't come up with her end of the financing. On the other hand Mrs. R. told me she had paid Riker $25,000 but was holding the final payment back until after the trip was over.

"The project bogged down right here. We paid the pilot $750 for his time and closed out the deal. There were other people waiting for the ship. It was the first of the big-wing special Broughams with cockpit in front and extra tankage under the fuselage — which made it look like a pregnant guppy.

"We sold it to an Oklahoma oil outfit that planned to use it for a record endurance flight. Red Harrigan test hopped X3598 with a full fuel load and turned it over to Al Henley, pilot for the oil company."

One report has it that the pot-bellied job was caught in a west Texas 'duster' and was destroyed. More likely the story concerns another Brougham whose pilot was apparently trying to fly 'on the deck' in near zero-zero conditions near San Angelo and went in on his back, rolling the plane up in a ball. The factory sent Harrigan and Fred Rohr down to see if there had been a structural malfunction, but none was found. They had hit a similar storm but landed and ran for cover.

The discrepancy in the story of *The Spirit of Holy City Love* is that one story has it that the plane was "due to be completed on March 20" while the San Diego Union of January 14 shows Al Henley taking delivery of the completed plane "which is now en route to Tulsa, Okla., for the endurance try."

We suspect that *Aviation* Magazine may inadvertently have run a Mahoney factory press release on Mrs. Rosencrantz projected flight several months late, causing the confusion in dates.

Tom Mathews \ H. A. Erickson

Al Henley, oil company pilot, planned record endurance flight in the Brougham

Ryan pilots Bill Monday, left and Doug Kelley, alongside the big-wing long-range Brougham

Bill Monday

Jerry Jones

Six B. F. Mahoney-built Broughams await delivery to customers at Dutch Flats field, San Diego

Enter the St. Louis Crowd

THE prominent St. Louis businessmen who had backed Lindbergh's trans-Atlantic flight couldn't have been more pleased with the recognition their air-minded city received. Their selection of the name *Spirit of St. Louis* had been fortuitous indeed.

How further to capitalize on the flight?

It was only a matter of weeks after Lindbergh had landed in Paris before one of his backers, Harold M. Bixby, president of the St. Louis Chamber of Commerce, was in touch with B. F. Mahoney. Would he consider moving the Mahoney plant to St. Louis?

For months nothing tangible came of the contacts made by Bixby and by Harry H. Knight, another backer. Then Philip DeCatesby Ball, aviation enthusiast, ice machine manufacturer and the gruff, rough-and-ready owner of the St. Louis Browns baseball club, became actively interested. In September he bought a Mahoney-built Ryan B-1 Brougham and sent his pilot, Frank T. Dunn, to San Diego to take delivery.

That same month, Robertson Aircraft Corp., operator of the St. Louis-Chicago Contract Air Mail route No. 2 for which Lindbergh had been a pilot, announced that they would build a factory at Lambert-St. Louis field. "It is impossible to get delivery of a new airplane for three to four months," Major William B. Robertson said. "The time is now right for the expansion of this branch of aviation." To emphasize his point Major Robertson announced the purchase of 1150 Curtiss OX-5 engines. Later he was to organize the Curtiss-Robertson company to manufacture the Curtiss "Robin," an OX-5 powered three-place monoplane.

While St. Louis aviation interests were geared to expansion, the Mahoney situation in San Diego was not so sanguine.

The company was on the ragged edge financially even before the Lindbergh flight. Frank Samuels of *Aero Digest* magazine had written his editors that "They are very short of money and must be recapitalized or close their doors."

And, as recalled by Tom Mathews:

"The company was on the verge of bankruptcy before Lindbergh, during Lindbergh and after Lindbergh. However, every employee got paid, somehow; every bill got paid eventually. But some of the transactions to keep the plant going were weird indeed.

"Wright Aeronautical got to the point where the Whirlwind engines they sent out had to be stored under conditions where they could be released only upon payment of $5,000 each at the time they were to be put in the Broughams. Mahoney had to keep borrowing from the bank, against the engines in storage, to meet the payroll which was running around $5,000 a week in those days. Sam Breder, the new sales manager who succeeded A. J. Edwards, had to keep collecting enough on advance plane orders to cover the loans against the engines.

"Mahoney may have lived the Roaring Twenties somewhat in the F. Scott Fitzgerald manner, but he was a pretty smart Irishman. He wasn't just a Ring-a-ding or a Rum-dum interested in gambling, drink and women. And he was generous, on occasion selling personally-held securities and depositing the receipts to cover payrolls.

"One Saturday afternoon Franklin wrote the payroll checks and then asked me how much cash I had from the day's sightseeing flights at the Dutch Flats field. I counted about $650. He said bring it along. We drove to Tijuana and he took it to the crap table and won about $2,500 in a couple of hours. He gave it to me to deposit first thing Monday morning to cover the payroll checks. This happened twice."

It was mid-October before stock certificates in the B. F. Mahoney Aircraft Corporation were available. Then all the management people were told to get busy and sell as many shares as they could, but the stock didn't move very fast.

Meantime, P. DeC Ball — together with Bixby and Knight — made an offer that Mahoney couldn't turn down. Reportedly they would pay $1,000,000 for the 5,000 shares, 55 percent of which were held by Mahoney personally.

"When the St. Louis crowd came in," Mathews recalls, "they wanted all of the stock certificates or no deal. Fortunately we had sold only about 400 shares and in a couple of weeks had them all back except *one* share I had sold to a bartender in Tijuana. After a couple of weeks of my leaning on his bar trying to get it back, Mahoney took over. He went down and finally bought that last share. That closed the deal with Ball and friends."

On the last day of 1927, Franklin Mahoney announced that his company would be moved from San Diego to St. Louis within 90 days and that a new Missouri corporation, Mahoney Aircraft Corp., would succeed the B. F. Mahoney Aircraft Corp. of California. He would remain as president with P. DeC Ball as Chairman and Knight and Bixby as the other members of the board of directors.

The St. Louis group agreed with Mahoney that immediately upon formation of the Missouri corporation they would "cause you to be elected as its chief executive for a period of five years at a yearly salary of $10,000 plus an annual expense allowance of $5,000 . . . and that you shall devote all of your time and energy to this business."

Within a period of six months three different corporations — Ryan Airlines, Mahoney and then the St. Louis group — had owned production rights to the Ryan Brougham. Three more 'Ryan' companies were to follow later as builders of the Brougham.

Vic Stuhr

Baseball magnate Phil DeC. Ball, an early Brougham customer, headed group which bought Mahoney's company. Vic Stuhr standing beside landing gear of Ball's plane

Ryan Aeronautical Library

Group of San Diego factory workers. Fred Magula, foreman, points to Mahoney name on rudder

Workmen finish fuselage assembly in plane-a-day production test

H. A. Erickson

Wing assembly department in old fish cannery building

Hugh Belknapp

A new factory was to be built immediately at Anglum, Missouri, adjacent to the new Lambert - St. Louis Field, but as Mahoney said optimistically, "The San Diego factory will continue to produce two Broughams per week and at the same time the factory in St. Louis will begin to produce an equal number.

"As soon as the St. Louis plant is running smothly we will transfer the San Diego factory to St. Louis 'en masse' and begin production of a Brougham a day."

In March, 1928 the recently selected vice-president and general manager of the new Mahoney Aircraft Corporation of Missouri arrived in San Diego to learn more about Brougham production and to prepare for the move to St. Louis. He was John C. Nulsen, former St. Louis manager of the National Malleable and Steel Castings Co. Not incidentally, he had been married to Phillipa Ball, the daughter of Philip DeCatesby Ball and mother of their son, but unfortunately she had died some ten years previously and Nulsen had remarried.

Nulsen soon came up with the idea of having a "Full Production Week" to ascertain factory capacity. Could the plant turn out a Brougham a day without increasing the size of the factory or work force?

Between April 16 and 21, six Broughams were assembled and test flown in a five and one-half day period. "The only thing that prevents plane-a-day production is lack of motors," announced Nulsen in a moment of excessive optimism.

* * *

As soon as Mahoney had made his deal with the St. Louis interests he cabled the news to Col. Charles A. Lindbergh then on a good-will tour of Central America with the *Spirit of St. Louis*:

JUST CONSUMMATED DEAL WHEREBY WILL MOVE PLANT TO ST. LOUIS NEXT NINETY DAYS. WILL MEET YOU IN ST. LOUIS UPON YOUR RETURN FROM CUBA. HAPPY NEW YEAR.

Tegucigalpa, Honduras
Jan. 4, 1928

Dear Franklin:

I am delighted to hear that you are moving to St. Louis, and that you will be there when I return from Cuba. I believe that St. Louis is unusually well located geographically for a plane factory and that you have made an excellent decision.

The "Spirit of St. Louis" is in perfect condition and performing as well as ever on this flight.

I have been considering placing this plane in the Smithsonian soon after returning to the States, although I do not want any publicity given until I am sure. If so I will need a plane to fly around in and am wondering if you have done anything along the lines we talked about in San Diego.

If Hall has nothing in view at present it might be best for me to take one of the regular models temporarily at least. Will talk to you about this in St. Louis.

With best wishes to all,

Charles A. Lindbergh

Courtesy Mrs. Samuel C. Breder

Along what lines had Mahoney and Lindbergh talked in September when the Lone Eagle had been in San Diego? Undoubtedly Franklin had told Lindbergh of his plan to furnish him a new Mahoney Brougham — to be a gift from all the manufacturers and suppliers whose equipment was used on the trans-Atlantic flight. The new plane for Lindbergh's own use would replace the *Spirit* which would go to the Smithsonian Institution.

The custom-built Lindbergh Special, later to be given a B-1X model designation, was to be a near duplicate of the *Spirit* with the same 46-foot span wing in place of the standard 42-foot span wing of the commercial B-1 Brougham. To avoid the instability which was characteristic of the *Spirit*, the tail surfaces would be increased in size. Again, the ailerons were set into the trailing edge some distance from the wing tip. Another feature, later found to be an undesirable one, was installation of landing lights in the leading edge of the wing, as in Frank Hawks' *Gold Bug*.

"We started this ship about February of 1928," recalls publicist Tom Mathews, "as a factory special with engine, instruments, accessories, etc., donated by the various companies who expected to receive credit and publicity for their contributions.

"The plane was to be presented to Lindbergh when he came out to San Diego in the spring. Art Mankey, the chief engineer, and I spent a lot of time nursing this job through the plant. The engine, a present from Wright Aeronautical, was a nickel-plated J-5 Whirlwind. It had an electric self starter; actually the usual Eclipse inertial starter with hand crank, but we had worked out a motor arrangement to wind up the inertial flywheel.

"There were many other special features in the luxuriously appointed B-1X, a number of them items like night parachute flares which Lindbergh suggested be added as we went along. One thing he didn't want was any publicity whatever about the new ship, although he was agreeable that pictures be taken of the actual delivery.

"Most of the suppliers were donating the original and extra equipment just for the publicity so I was in a real bind. Mahoney told me to go along with Lindbergh's request, but that it was okey to take pictures of the various equipment items 'for the files.'

Expediting special job through factory are Art Mankey, far left, and Tom Mathews, far right

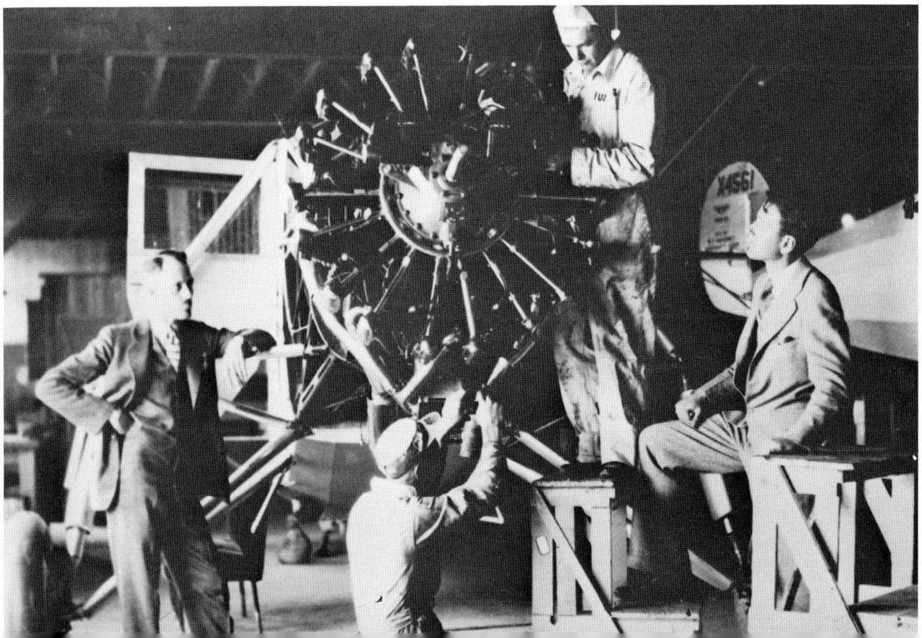

Ryan Aeronautical Library

"Finally on April 5, 1928, Lindbergh took delivery of NX 4215. It was the company's serial number (c/n) 69. Red Harrigan had test flown it some days earlier and demonstrated quite a few tricks such as a nose-high power slip, directional changes by opening and closing the doors (in case the rudder jammed), and rudder turns with dead ailerons. Art Mankey and I were with Red when he tried that nose-high power slip over the electric pole wires at the end of the field. It had to be the shortest, brake-less landing on record!"

Don Hall, who was working on his new experimental plane and had little to do with the Lindbergh Special, was less impressed. On Lindbergh's first flight in the B-1X, Hall rode along as an observer but was not satisfied with its performance. Nevertheless, Lindbergh decided to accept it.

This less-than-enthusiastic acceptance was somewhat at variance with the press reports about the new plane's first flight. One story reported it this way —

"After a preliminary test flight with Don Hall, Lindbergh went over to J. J. Harrigan, test pilot for Mahoney, and the two conversed in low tones for several minutes. Then with a broad smile on his lips, Lindbergh invited chief engineer W. A. Mankey and B. F. Mahoney to take a hop with him in his new plane.

"At a maximum altitude of fifty feet he throttled back the motor and turned around in his seat. Hands and feet off the controls he began making low turns by opening and closing the right hand door of the Brougham. When he motioned Mankey to open and close the left hand door, the plane turned in the other direction.

"Mahoney and Mankey showed their uneasiness so Lindbergh headed back for the field and made a perfect landing. A moment later Lindbergh said to Mankey, 'It's just right, I like it.' "

No doubt Lindbergh didn't like the impression Tom Mathews' publicity yarn left with readers nor the special advertising section on the B-1X which appeared in the May issue of *Aero Digest* magazine.

The day after Lindbergh's test flight, a crowd had gathered at Mahoney (formerly Ryan) Field to see the plane take off. At a sign from Lindbergh his four guest passengers, Harry F. Guggenheim, and three St. Louis backers — Knight, Bixby and William B. Robertson — ran out to the plane and got aboard.

After spending a week at a ranch near Santa Barbara, Lindbergh took off in the B-1X for St. Louis on April 13 with Knight and Bixby as passengers.

Lindbergh expected to make an overnight stop at the Grand Canyon. Writing a friend years later he explained that "I had planned on landing on a field that

Charles A. Lindbergh and his B-1X special Brougham which replaced his "Spirit of St. Louis"

St. Louis Globe-Democrat

had been described to me as lying not many miles from the El Tovar Hotel on the south rim of the canyon. We arrived over the area in early evening, and I could not find the field. I decided to continue south to Williams, where I knew a satisfactory landing area existed.

"Since I had planned on landing well before nightfall, I had not checked to see that the parachute flares were releasable. (It was the practice to 'safety' the bottom of the flares with cotter-pins for daylight flights.) These cotter-pins were reachable by a man laying on the bottom of the cabin.

"I told Bixby and Knight that somebody would have to check on the cotter-pins, and pull them out if they were in place. Actually, it wasn't either difficult or dangerous; but a good deal of humorous discussion resulted. Finally, Bixby held on to Knight's legs while he pushed through the cracked-open cabin door, bent under the fuselage, and pulled out the cotter-pins with a pair of pliers.

"When we circled over Williams, a number of cars drove out and threw their headlights over the grass-field area. I landed without using either of the flares.

"A later newspaper report (the next day after taking off for St. Louis) that I had landed on a country road six miles from the airport because of propeller trouble is pure fiction.

"I used the new Brougham on many flights, transcontinental and others, and carried a good many passengers in it. As I recall I used it on all of my early survey flights while laying out the Transcontinental Air Transport passenger route between New York and Los Angeles. I eventually returned the Brougham, in good condition to the B. F. Mahoney company."

It was a distinct disadvantage having an X-licensed airplane particularly since the word was fairly widespread that Lindbergh was not too happy and wanted to dispose of it. In October Art Mankey, by then in St. Louis as chief engineer, was writing to Henry Breckinridge, Lindbergh's attorney, to advise that "this ship can now be licensed permanently under the B-2 type class." This would allow issuance of an unrestricted license and permit its resale as a commercial plane.

* * *

No sooner had Lindbergh departed San Diego in his new Mahoney B-1X Special, NX4215, than Captain Emilio Carranza of the Mexican Army arrived at the factory to inspect his new long-range Brougham.

Spurred by *Excelsior,* Mexico's largest daily newspaper, purchase of the plane was financed by popular subscription to show the friendly and respectful admiration of the Mexican people for Lindbergh. Carranza would make a non-stop Mexico - Washington flight as a good-will gesture, repaying Lindbergh's December visit to Mexico.

Just 22 — younger than Lindbergh — Carranza was a product of the U.S. Army flight training school at San Antonio, Texas. A great-nephew of President Carranza, he was already the best-known aviator south of the Rio Grande.

Like the *Spirit,* the big-wing *Excelsior* carried hundreds of gallons of fuel, mostly in fuselage tanks, but the pilot was

Missouri Historical Society

Lindbergh with B-1X at Grand Canyon Airport with backers Harold M. Bixby, left and Harry H. Knight, right

Ryan Aeronautical Library

Capt. Emilio Carranza, right, with sales manager Sam C. Breder

H. A. Erickson

Carranza good-will flight was sponsored by Mexico City newspaper, Excelsior

H. A. Erickson

Emilio Carranza in the cockpit

placed forward immediately behind the engine as in the standard commercial version of the Brougham. One thing Carranza faced that Lindbergh did not have to contend with — the heavily-laden *Excelsior* would have to take off from the 7,600-foot altitude Valbuena Airport at Mexico City!

May 24, 1928 — almost precisely a year after Lindbergh flew the Atlantic — Carranza made it non-stop from Camp Kearney in San Diego to Valbuena Airport at Mexico City in 18 hours, 20 minutes.

Three weeks later, June 11, after a 2,500-foot take off run because of its 326-gallon fuel load, the *Excelsior* rose from the Valbuena strip at 8:10 a.m. and headed for Washington, D.C.

After 18 hours, 35 minutes, during which he battled rain, storms and fog, and had been sighted over Tampico, Brownsville, New Orleans, Montgomery and Atlanta, Capt. Carranza was forced to make a 4 a.m. unscheduled landing at Mooresville, North Carolina. It was an emergency field on the New York - Atlanta airmail route.

The next day Carranza flew on to Washington, and later to New York. The forced landing in North Carolina was diplomatically de-emphasized in the warm flurry of goodwill speeches and parades that welcomed the Mexican pilot to the United States.

Nevertheless, Carranza had failed to carry out his mission as planned, but he intended to redress the failure by flying home next month non-stop from New York to Mexico City.

July 13 was a day of thunderstorms. The young Mexican officer waited for one to let up, then headed south. There was, however, another severe storm over the Pine Barrens of southern New Jersey. After being in the air less than an hour, Carranza and his heavily-laden plane were in trouble. Above the rumbling thunder, residents of Chatsworth heard the plane but since no crash was seen or heard no search was started.

Next day, Henry Carr and his wife, out gathering wild blueberries, came upon the wreckage of the *Excelsior*. Carranza's body was carried out and returned to Mexico City, where three hundred thousand people followed the cortege to the Dolores Cemetery.

Every July since then a ceremony takes place simultaneously at Carranza's tomb in Mexico City and at the isolated Carranza Memorial in the wild Pine Barrens.

An Aztec falling eagle stands out in relief on one side of the monument, and Spanish words spill down another in memory of "capitan aviador Emilio Carranza, muerto tragicamente el 13 de julio 1928."

Ryan Aeronautical Library

Taking off the morning after a night forced landing in North Carolina

Ed Morrow

Some of the last B-1 Broughams to be produced in San Diego await delivery

Now It's Mahoney-Ryan

By mid-summer 1928 Mahoney was still building Ryan Broughams in San Diego, having announced earlier in the year that the company would retain its west coast plant at least one more year. According to John Nulsen it was planned to increase production and the working force at San Diego since the new $300,000 plant at St. Louis would not be ready until Fall.

Under new sales manager Sam Breder's direction, additional domestic distributors and dealers were being signed up, and export marketing was expanded. During a visit to Pittsburgh when a new east coast distributor was signed up, Mahoney stated the company might also establish a factory there.

There was no question but what the Brougham was the market leader in the broadening commercial aircraft industry and that Lindbergh's magnificant flight had touched off an unprecedented period of expansion. *Aero Digest* magazine wrote –

> " . . . almost overnight the world became airminded. The barriers of disbelief, the bugaboo of fear, the too skeptical attitude of capital which had retarded industrial growth of aviation for years were suddenly swept aside. One magnificent flight had accomplished in 33½ hours what hundreds of men had been trying to do for years."

Time and again articles about the expansion of aviation were illustrated by pictures of Lindbergh and Mahoney standing beside the Ryan Brougham.

Clearly the Mahoney organization was riding the crest of popular acceptance of flying. Reportedly Mahoney used more of the commercial output of the Wright engine factory than any other aircraft plant. No less than four and sometimes six Broughams a week were being produced.

Business organizations all over the country – oil companies, newspapers, airlines, ranchers, exporters, sales executives, government officials, promoters, aircraft distributors and dealers – were ordering aircraft in unprecedented numbers and a large percentage of them were Broughams. But to the buying public they were Ryan Broughams – not Mahoney Broughams.

This was no doubt the motivation behind yet another name change – this one restoring the name Ryan to the company manufacturing the Brougham. Mahoney-Ryan Aircraft Corporation, formed July 1, 1928 and capitalized at $1,750,000, carefully identified the firm as "Owners of Ryan Airlines and Ryan

Guatemalan fliers stop in St. Louis during good-will flight to Washington, D.C.

Ryan Aeronautical Library

H. A. Erickson

Miguel Zuniga cultivated the Latin American market. In cockpit, Doug Kelley

Flying Company." These were, of course the operating names of T. Claude Ryan's original businesses established in San Diego six years earlier. However, Claude Ryan was in no way affiliated with the re-named company, having started anew in aviation in San Diego after Lindbergh's flight under the name Ryan Aeronautical Corporation. At that time Claude's old friend, Hawley Bowlus, left Mahoney as factory superintendent to rejoin his former boss.

Under Nulsen as general manager and A. G. (Bert) Tindale, a mild-mannered Australian, as factory superintendent, production at the abandoned fish cannery plant in San Diego increased in efficiency. With the opening of the adjacent Lindbergh Field in August, the company no longer had to truck planes to the old Mahoney Field on Barnett Avenue for final assembly.

Wings went straight through on the second floor of the plant and fuselages on the first floor, to be joined together at the end of the production line and flown away from the factory doors. The proximity of San Diego Bay was also a great convenience so that seaplane versions of the Brougham, which had just been developed, were easily rolled to the water's edge on the other side of the factory for testing.

Latin America proved to be a particularly lucrative field for sale of Ryan Broughams under the knowledgeable efforts of M.A. (Mike) Zuniga, diplomat turned salesman, who had been in on the Carranza *Excelsior* deal.

For the first time Central American governments turned from an exclusive interest in military aviation to the potential of commercial air transportation. President Chaco of Guatemala purchased three Broughams to inaugurate airmail, passenger and express service from Guatemala City to isolated provinces. Typical of the services inaugurated late in 1928 was that to Peten in northern Guatemala. The three hour flight supple-

Ev Cassagners

Washington-New York Airline started service with Broughams in 1929

mented the ten to twenty days travel by narrow-gauge railroad and pack-trail through British Honduras and Mexico. The following year Pickwick Airways was to establish scheduled service between Los Angeles, Mexico City and Guatemala City with a later version of the Brougham, the B-5 model.

While Mike Zuniga cultivated the Latin American market, Brougham owner Count Giovanni Bonmartini took on sales representation at Rome, Italy.

New ads, again featuring the Ryan name, began appearing in the trade press. "No job," they read, "is too big for Ryan Brougham.

"Vance Breese stepped into a standard Ryan B-1 Brougham on the morning of August 8th and, with a full load of passengers in c/n 118 flew over Pikes Peak at an altitude of 17,000 feet. The next day he carried camera equipment in addition to four passengers and again flew over the famous summit.

"Some said it couldn't be done with a cabin plane, but Ryan Broughams are delivering the Los Angeles Times to San Francisco and Oakland, leaving after midnight, arriving at dawn with 800 pounds of newspapers packed in the cabin behind the pilot. . . " As a clumsily contrived afterthough they added . . . "on express train schedule, every night."

* * *

For years, accomplishments of the B-1 Broughams continued to make news, but for every ill-starred publicity stunt flight like that of "Leo the Lion" or the "Holy City to Holy City" fiasco there were 150 other Broughams performing useful work-a-day tasks.

Still, many Broughams got into the news with more positive stories about aviation.

Frank Hawks had taken delivery of the *Gold Bug* — the first B-1 — in time for his c/n 29 to be the only Brougham entered in the 1927 Ford Reliability Tour. By the next year three Broughams were entered in the National Air Tour.

Al Henley, piloting the Texas Pacific Coal & Oil Company B-1 (5547, c/n 112) placed seventh in 1928; and Vance Breese, flying NC5553 (c/n 118) for Mutual Aircraft of Los Angeles, the largest Brougham distributor, came in ninth. In qualifying to participate, Henley had demonstrated he could take off in ten seconds and complete his landing from touchdown to complete stop in 4½ seconds.

Way back in 20th place was the ever-popular E. W. "Pop" Cleveland flying NC1159, c/n 38, one of the earliest B-1s built and the first to be equipped with Cleveland "Aerol" (air-oil) struts for the landing gear.

Ryan Aeronautical Library

George B. Fredell, oil executive, left, with Mahoney-Ryan sales manager, Sam C. Breder

Brougham c/n 79 was a big-wing B-1 with 46-foot span wing and large tail surface, licensed as X4654 and purchased by Hall and Briscoe of Chickasha, Oklahoma. Joe Hart, Jr., a partner in that firm, used it in a May 1928 flight attempting to set a solo endurance record. He had to give up the flight in the 'X' licensed plane after 9 hours 38 minutes Hart was short of oil because at the high RPM he was using he had burned up the piston rings.

The plane was then sold to Paul R. Braniff and became Ship No. 2 of Braniff Airways as they inaugurated service between Oklahoma City and Tulsa late in June 1928. Among the incorporators of the airline were F. C. Hall and Powell Briscoe.

A new engine was installed, the extra gas tanks removed from the passenger compartment and the upholstery and seats replaced. Braniff advised the Department of Commerce that "I am removing the letter 'X' from Ryan Brougham #79 and upon receipt of the new license card will add the letter 'C'. In the meantime I am using this ship for the hauling of passengers on the Oklahoma City - Tulsa air line whenever necessary."

A month later Braniff wrote the Department of Commerce that to make it feasible to use the plane in scheduled passenger service "it will be necessary to cut the wings down to the standard 42-foot size. We will not interfere with ailerons, but will simply take a certain amount off the end of each wing and will also put on the standard tail group.

"We are very anxious to use this ship immediately on the Ponca City - Tulsa passenger line and if this work is done by Government-licensed mechanics and reported to you will it be necessary to have it re-inspected before we could operate it as a Government licensed aircraft?"

Things apparently moved slowly be-

The ever-popular "Pop" Cleveland was a regular at all aviation events

Ev Cassagners

Ryan Aeronautical Library

Len Morgan

Hall and Briscoe big-wing Brougham became Ship No. 2 of Braniff Airline

cause three months later the Department of Commerce representative at Tulsa was recommending to Washington that "the plane be grounded because Braniff was remodeling without reinspection and carrying pay passengers cross-country . . . " In due course NC4654 was finally licensed, but airline enforcement procedures have materially changed in the past 45 years.

A B-1 with license 4474 (c/n 72) was used for tests of the seaplane configuration with twin Edo pontoons. Carrying 16 fewer gallons of fuel than the B-1 landplane, to compensate for the added weight of the floats, the seaplane version was nonetheless able to take off with pilot and four passengers in 18 seconds.

"High location of the Brougham wing," read an Edo ad, "allows the seaplane to be brought sideways to any moderately high dock, pier or barge like a motor boat." Herb Faye then a pilot for Capitol Airways, Brougham distributor at Washington, D.C., appears to have operated 4474 on the Potomoc River at the nation's capitol.

The west coast plant of Mahoney Aircraft being located adjacent to San Diego Bay, tests of a seaplane version of B-1, license 6654, c/n 139, were also conducted there in July 1928 with Red Harrigan, chief pilot, at the controls.

"The tests," says Tom Mathews, "were run with full load — pilot and four passengers and full gas. No problems except the take-off was a little longer than I had anticipated and Harrigan was just getting up on the 'step' as he passed the launch I had hired for picture-taking and to cruise the course before takeoff."

W. H. (Hal) Wolfe of Manitoba Basin Mining, Limited, used B-1 license 4450 in mining operations out of The Pas, Manitoba, some 250 miles northwest of Winnipeg.

Another B-1 — NC4559, c/n 73 — was, like the *Queen of the Yukon,* a ski-equipped Brougham, which operated in Alaska for many years for Bowman Airways.

After his barnstorming tour with the *Queen of the Yukon,* Ryan pilot Bill

Bill Monday

Herb Fahy flew pontoon-equipped Brougham on Potomac River in Washington, D.C.

Howard Bowman

S. R. Winters

Bowman Airways, left, put its Brougham on skis in Alaska. Kingston-Rhodes Airways lashed canoes to pontoons and took off on hunting trips

Monday returned to San Diego where several of his Brougham flights made news.

The Mahoney sales people felt there would be a real advantage in setting an altitude record for stock commercial planes with the B-1. Tom Mathews, the company's ingenious publicist, lined up his friend Max Miller of the San Diego Sun (later the famed author of "I Cover the Waterfront.") to accompany Monday on such a flight.

Taking off from Mahoney Field on Dutch Flats soon after noon, Monday pointed the nose skyward. In an hour and 25 minutes the Brougham would climb no higher and Monday began the rapid descent. Two barometers had been carried aloft. One broke at 20,000 feet; the other continued to operate and when checked after landing showed a reading slightly in excess of 24,000 feet.

Monday had made the flight without benefit of oxygen. Miller had passed out — gone to sleep — in the rarified atmosphere at 20,000 feet.

Some months later, Exclusive Florists of San Diego chartered a Brougham and engaged Monday to fly owner Graham Dible and Chamber of Commerce executive D. W. Campbell to the east coast with a cargo of orchids valued at more than $1,000. It was the start of air shipments of fresh flowers from San Diego County, a major source of air freight revenues from the west coast for the past four decades.

In the South Pacific, B-1 Brougham c/n 148 made local history in November 1928 when its Australian owner, Capt. Hurley of Sydney, and two companions safely crossed the Tasman Sea to New Zealand for the first such trip on a 1,339-mile flight. And in New Guinea, the ability of the Brougham to withstand the operational rigors of high heat and humidity in a remote tropical area was demonstrated.

Among the tuna boats in San Diego Harbor c/n 139 makes its debut as a seaplane

H. A. Erickson

Albert Tindall, Mahoney-Ryan Factory Superintendent

W. Arthur Mankey, Chief Engineer of the Mahoney-Ryan Corp.

John C. Harrigan, Chief Test Pilot of the Mahoney-Ryan Corp.

Sam C. Breder, Mahoney-Ryan Sales Manager

Robert B. D. Adams, in charge of Advertising and Publicity

Strauss-St. Louis

John C. Nulsen
Vice President and General Manager

B-3s, B-5s and Doodle Bugs

With availability of the newest 9-cylinder Wright Whirlwind engine, the 300 h.p. J-6 model, Art Mankey and his engineering staff started work in November on a six-place version of the Brougham which was later to have the designation B-5.

Meanwhile during a transition between the B-1 and B-5 models a number of modifications were incorporated in the Brougham including a wider, larger cabin interior with optional dual control provision for two pilots. Larger tail surfaces provided better directional control and there were other design improvements including moving the ailerons inboard from the wing tips as on the 46-foot span versions. Eight such B-3 models, c/n 179-186, were built, and Approved Type Certificate No. 104 was issued January 24, 1929 for the six-place J-5 powered model.

While the eight B-3s came out with the 220 h.p. J-5 engine, a number were converted to test the larger J-6 300 h.p. Whirlwind installation which was to become standard on the B-5 model.

By fall of 1928, the decision had been made to move all production to the new St. Louis factory, but only a handful of factory employees were willing to cast their lot with the new owners and put down roots in the mid-west – or as they saw it, the mid-east.

Key executives, however, did move to St. Louis and take over management responsibilities there. From San Diego, in addition to Franklin Mahoney, as president, came Art Mankey, chief engineer; Sam C. Breder, sales manager; Bert Tindale, factory superintendent; Karl F. Lueder, purchasing agent and J. J. (Red) Harrigan, chief test pilot. Veteran pilot Bill Monday also signed on but didn't remain long.

Three key factory employees, not wanting to go to St. Louis, struck out for themselves in Los Angeles as the Ryan Mechanics Monoplane Co. Their story is covered elsewhere.

The irrepressible but not irreplaceable Tom Mathews went east but soon left St. Louis for greener pastures and his advertising and publicity chores were taken over in November by B. D. (Bob) Adams. Dynamic sales manager Sam Breder, who unfortunately disliked planes and hated to fly, surfaced briefly in St. Louis then returned to California where the climate was more to his liking.

Bob Adams, a young Hollywood writer, had signed on in June to help Tom Mathews answer fan mail, write advertising copy, and show customers around the plant and entertain them while their planes were being finished. Adams had been inspired to get into aviation during a stint at Universal Studio in which a sleek Ryan M-1 monoplane was flown in a thriller, titled "The Flying Cop," to chase the bad guys trying to escape in their obsolete Jenny biplanes.

In September, Adams left San Diego on a southern sales trip and ended up before Christmas 1928 in St. Louis as assistant sales manager and head of advertising.

Meanwhile, on October 5, 1928, wings for the last Whirlwind-powered Ryan B-1 Brougham built in San Diego were completed. It was Serial No. 178. The last five employees closed the fish cannery plant October 19th.

Six Broughams carrying a dozen workmen made the San Diego - St. Louis flight as the transfer got under way. John C. Nulsen, vice president and general manager, reported that the men went right to work assembling fuselages, and that the first St. Louis-built Brougham would be completed by November 15.

* * *

When the production equipment and inventory of Brougham parts was sent from San Diego to the new St. Louis plant, the shipment also contained one other plane — Don Hall's experimental X-1 Mahoney-Ryan Special.

Don and his assistant, Wilhelm F. (Bill) Schult, had begun design work on the radical two-place monoplane in November 1927. Mahoney and Hall had agreed on the general configuration of the plane and they hoped to build them by the hundreds for the surging training and private plane market opened up by Lindbergh's flight to Paris. By September 1928 they were ready for the first test flights at San Diego.

The most radical change was the absence of fixed tail surfaces — both stabilizer and fin. The X-1 was said to be the first monoplane in which a movable 'stabilator' — virtually an aft wing with lifting characteristics—replaced the usual stabilizer-elevator combination. The plane also had an extra-large rudder which, as with the stabilator, pivoted like a weathervane.

Other features included five degrees dihedral in the very short wing, a wide landing gear and a high static angle on the ground so that "in effect, the plane lands on the wing and not the landing gear."

While the X-1 had conventional stick and rudder controls some extra features were added, principally a "gear shift" with 12 positions to vary the leverage the stick could apply to the lifting stabilator. A shift from one gear to another greatly affected the X-1's handling characteristics, changing the plane to a 'trick' ship capable of unusual maneuvers.

In early tests Hall also wanted to check the ability of the unique wing airfoil to handle large center of gravity shifts. To prove his theory, a temporary track with a 50-pound weight on a dolly was installed to run the length of the fuselage. Later it was removed.

Tom Mathews describes the first flight:

The plane was taken out to Kearny Mesa (where Lindbergh's *Spirit* was tested) and Red Harrigan, chief pilot, got some ground time with the tail propped up in level position and the 90 h.p. Warner radial engine at cruise throttle.

Ex-Navy pilot Harrigan was qualified in flying boats, pontoon boats, balloons, dirigibles, fighters, bombers, transports and utility planes but he was the first to admit that his first flight in X7621 was completely unrelated to any previous flying experience.

John W. Underwood

Actor Neil Hamilton, center, flys a Ryan M-1 in the role of "The Flying Cop".

Wilhelm Schult

Wilhelm Schult poses with Don Hall's X-1 experimental monoplane

Mahoney-Ryan Special had a unique moveable aft wing in place of the usual stabilizer-elevator combination

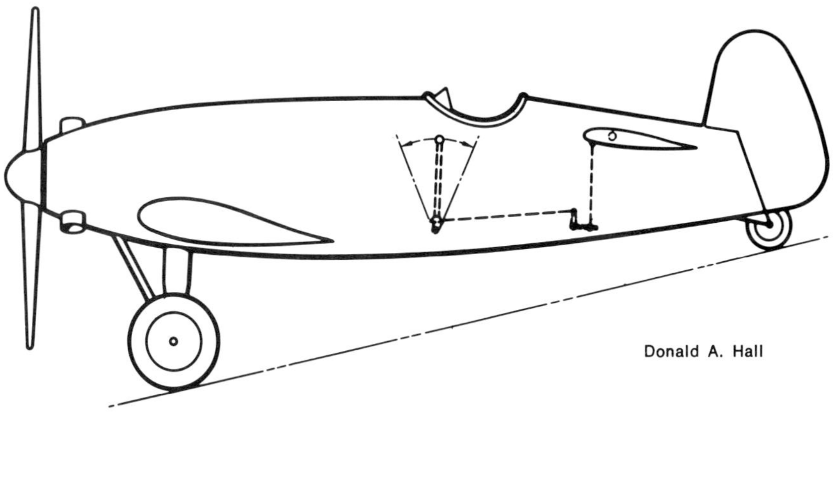

Donald A. Hall

Tom Mathews

. . . Red Harrigan instinctively shifted into low gear and thus had no control . . .

B. D. Adams

At St. Louis, Charles A. Lindbergh, right, made several flights in Hall's X-1 special under watchful eye of Red Harrigan, left.

On take-off the tail kept getting higher than the nose. In its first turn the plane crabbed sideways and started to slip toward the ground. For a while it appeared the plane and pilot would disappear over the horizon bouncing like a kangaroo. Harrigan finally got the ship headed back to the field and landed.

It was then Hall explained the function of the "gear shift" lever. Harrigan had been flying the plane with the lever in the last notch — the 12th position!

The flight tests continued and as Harrigan gradually taught himself to fly the plane and gained confidence he manipulated the gear shift, and weight changes fore and aft, under good level flight control, made sweeping turns over the field at 135 mph and completed a variety of landings.

Harrigan refused to spin the plane because, as he explained to Mahoney, he wasn't sure that X7621 wouldn't spin tail first "and I don't want to be the first guy in the world to find out about that."

Hall's explanation of the first flight was, understandably, somewhat different than Mathews' report. Wiring Lindbergh, he explained:

RED (HARRIGAN) GOT INTO WHIP-STALL IN FOG TWO HUNDRED FEET ABOVE GROUND. THOUGHT HE WAS SURELY GONE BUT SHIP SLOWLY FELL OFF ON ONE WING, NOSING DOWN GENTLY. RED SAID NO OTHER SHIP COULD HAVE COME OUT AT THAT ALTITUDE.

A week later from his room at the YMCA, Don Hall wrote "Slim" in more detail.

"In the whipstall that Red got into he had instinctively shifted into low gear and thus had no control. Yet the ship had the unusual stability characteristics of gently coming out to a controllable attitude of flight. There are 12 variable speeds in the leverage between the stick and the elevator, low gear being for cross-country flying only, giving small elevator movements.

"Certain important changes and improvements are now being made on the ship."

The Hall Special was reassembled in St. Louis and more tests run, some with a 110 h.p. Warner Scarab engine installed. Under Harrigan's counseling Charles A. Lindbergh made several flights in X7621 and reportedly thought it was pretty touchy.

B. D. (Bob) Adams, who succeeded Tom Mathews as Mahoney-Ryan publicist, recalls that "in Spring 1929 I undertook to fix it up with the help of some of our mechancis and actually made a straight-line flight with it one evening, skimming the full length of Lambert Field.

H. A. Erickson

Claude Ryan in 1929 purchased Mahoney operations and again had all flight activities at Dutch Flats

"Phil Ball, however, was afraid I would kill myself and he had his pilot, Frank Dunn, take me up a few days later and really wring the plane out. It was the first and only time flying ever made me sick."

The Mahoney-Ryan Hall Special was not the answer to the company's need for a broader product line despite a December 1928 announcement that "after some further refinements, this model will probably be placed in production at the Anglum plant." As Mathews explained, "The company, $50,000 wiser, dropped the whole project as having more problems than possibilties."

* * *

Although the manufacturing operations in San Diego shut down in Ocober 1928, the flight training and aerial service activities of the Mahoney organization continued until April 1929. Then on April 11 the T. C. Ryan Aeronautical Corporation, wholly owned by Claude Ryan, announced it had purchased the remaining San Diego activities of Mahoney. Ryan had bought the holdings and business through a third party, C. C. Edgington, who had purchased the properties a month or so earlier from the St. Louis owners.

In a flurry of activity that Fall of '28 at the new St. Louis plant, the first "Brougham for Six" (a B-3 in six-place configuration) was rushed to completion for display along with a B-1 at the Chicago Air Show in December. Two Broughams, in fact, were being advertised — the B-3 with J-5 engine as a five-place plane and the new B-5 with J-6 Whirlwind engine for six. The latter went into regular production in March 1929.

"The Mahoney-Ryan Brougham" the advertisements read, "takes off faster and lands slower, with its full load, then any other plane of its type, and we build and sell more Whirlwind cabin monoplanes than any other maker. This volume makes possible the attractive price."

The price may have been attractive, and the volume relatively high, but somehow, somewhere, something must have been missing in the economic equation. The stock market was still ten months short of the disasterous crash to come in October 1929, but Mahoney-Ryan was either in trouble or its St. Louis owners were getting anxious to bail out.

Franklin Mahoney, nominally president, had gone to Europe and executive decisions were in the hands of Phil DeC Ball, board chairman, who in turn had delegated operating authority to his son-in-law, John C. Nulsen, as executive vice president and general manager.

Near the close of 1928 the *St. Louis Post-Dispatch* headlined the story of yet another upheaval in the ever-changing fortunes of Mahoney's company.

B. F. MAHONEY QUITS
AIRPLANE COMPANY
Sells All Holdings in
Manufacturing Concern
St. Louisans Control

Gerry Balzer

One of the first St. Louis-built Broughams was this B-3 model, later donated to endurance fliers Robbins and Kelly

Wilhelm Schult

New Mahoney-Ryan St. Louis plant is background for one of eight B-3 Broughams built

Jerry Jones

Reg Robbins and Jim Kelly set new endurance record in second-hand B-1 Brougham

Ryan Aeronautical Library

New women's altitude record was established in B-3 Brougham by aviatrix Marvel Crosson

Mahoney's resignation as president, the article revealed, had been accepted a week earlier while he was in Europe, and he had sold all his stock in the company. No reason for Mahoney's disassociation with the firm was given.

While Mahoney had 'fronted' for the company in San Diego and later in St. Louis he was never actively in charge of either day-to-day operations or longer-range planning. And, unfortunately, Ball and his son-in-law were often not in agreement.

"The relationship," Bob Adams recalls, "was fine as long as John was on the west coast and Phil Ball didn't look in every day. But as soon as we got established in St. Louis, Ball had his chauffeur drive him, from his home nearby, to the plant first thing every morning on his way downtown. He would stay a couple of hours, mostly in the factory.

"Ball had a few favorites who were his source of plant gossip and complaints, and he would later chide Nulsen about these. The relationship deterioriated throughout 1929. Ball called the signals and Nulsen ran the plant the best he could with Ball's interference. Nulsen was a good manager in the sense of keeping things in balance but knew little of flying and was not a marketing man. He was a good, fair-minded man but in the wrong job."

* * *

Not only were the Broughams selling better than any other cabin plane of its day, they were also continuing to make news in highly publicized record flights.

In January 1929 the Army Air Corps had made headlines when a fiveman crew headed by Major Carl Spaatz and Captain Ira C. Eaker set a new refueling endurance record of 150 hours 41 minutes in the *Question Mark*. A brash young Texan, Reginald L. (Reg) Robbins, though he could beat that record with a two-year-old Ryan B-1 Brougham.

The plane, c/n 52, license R1766, had been delivered in November 1927, and had already logged 525 hours on its 220 horsepower J-5 engine when Robbins and co-pilot/mechanic 'Cowboy' James Kelly took off from Meacham Field in the *Fort Worth* just before noon on Sunday, May 19, 1929, carrying some 250 gallons of gas.

Early Monday morning another Brougham, the *Midland* flown by K. K. Hoffman of Tampico, Mexico, and H. S. Jones, of Midland, Texas, transferred 110 gallons of fuel through a 37-foot hose to the *Fort Worth*. Through daylight and darkness Robbins' and Kelly's plane droned on for six days, interrupted only by the twice-a-day refueling contacts and the occasional hazardous 'EVA' — extra vehicular activity — trips along the catwalk to oil the rocker arms of the J-5 engine.

On the second day, the metal buckle of Jim Kelly's safety belt accidentally struck the whirling wood propeller. Soon the dent became a crack and set up a serious, continuous vibration. Rain caused the crack to split further until it extended to 15 inches. Although finding themselves far short of the record, the two pilots determined to stick it out, finally landing on May 25 with a new sustained flight mark of 172 hours and 32 minutes. There were 17 refuelings, and a total of 1,510 gallons of gasoline went down the hose to the endurance ship while in 'orbit' over Fort Worth.

By coincidence, the second anniversary of Lindbergh's New York-Paris flight occurred during the Robbins and Kelly endurance flight.

Phil Ball was so pleased with the performance of the Brougham that he announced Mahoney-Ryan and Wright Aeronautical — prodded by Amon G. Carter, publisher of the *Fort Worth Record Telegram* — would make a gift of a brand new Brougham and engine to Robbins and Kelly. It was c/n 184, license NC7736, one of the B-3s converted to use the 300 h.p. J-6 Whirlwind engine. Reportedly they also collected $50,000 in prize money.

Two days after the *Fort Worth* had landed, 20-year-old aviatrix Marvel Crosson, sister of famed Alaskan bush pilot Joe Crosson, took off from Los Angeles in another of the B-3 Broughams powered with the 300 h.p. J-6 engine (c/n 185, license NC7737).

When she landed she held the new world altitude record for women with a flight to 23,996 feet, exceeding the previous mark by 3,726 feet. The flight had been sponsored by Mutual Aircraft, one of the largest Mahoney-Ryan distributors.

Zane S. Cieslak

First Brougham built under new "Ryan Aircraft" name was B-5 c/n 218 completed August 27, 1929

Detroit Aircraft Buys In

In May, 1929 the word began to get out that Mahoney-Ryan was to be included in a new holding company being put together in Detroit to offer a complete line of commerical and military land, sea and amphibian planes and lighter-than-air craft.

Firms mentioned as also being in the new combine included Aircraft Development Corporation and an American affiliate of Blackburn Aircraft of England with Western Hemisphere manufacturing rights. The 'leading light,' with grandiose ideas of trying to create a 'General Motors of the Air' and make Detroit the aircraft as well as automotive capital, was Edward S. Evans of the National Gliding Association and the Evans Auto Loading Company but the big names behind him included top executives of the auto industry.

Instead of the originally proposed name – Continental Aircraft Corporation – the new combine chose Detroit Aircraft Corp. as the name of the $20 million corporation, and soon had picked up not only Mahoney-Ryan and the other two original firms but also Lockheed Aircraft Corp. and eight other companies, including Parks Air College of East St. Louis.

The initial step in the transaction was a bookkeeping consolidation of Mahoney-Ryan and Aircraft Development Corporation, maker of metalclad dirigibles. Both were then merged into Detroit Aircraft. Old stock certificates show there was an exchange of Mahoney-Ryan shares for 200,000 shares of Aircraft Development stock.

In turn, another corporation was formed as a wholly-owned subsidiary of Detroit Aircraft to continue the Brougham operation in St. Louis. Paradoxically it was called Ryan Aircraft Corporation as the name Mahoney was completely dropped. The St. Louis stockholders came out with $2 million but to their later regret 90 percent of it was in the form of stock in Detroit Aircraft Corp.

The Detroit Aircraft board of directors was literally a who's who of the automotive industry – Charles F. Kettering and Charles S. Mott of General Motors; William B. Mayo of Ford auto and Ford trimotor aircraft fame, Roy D. Chapin of Hudson, and R. E. Olds of Reo. Also serving on the board were the St. Louis interests – Ball, Knight and Bixby – and two executives of Northwest Airways.

Foreign investors were attracted by the potential of a 'General Motors of the

B. D. Adams

Lindbergh test flies a new Brougham at St. Louis. Left to right: Bob Adams, pilot Buddy Jones, Phil DeC. Ball, Lindbergh, Red Harrigan

Air' and a syndicate in The Netherlands reportedly purchased over a million shares of Detroit Aircraft stock.

Ryan Aircraft would continue production in its new plant at Anglum, Missouri, on the edge of St. Louis-Lambert Field and John Nulsen would be retained as general manager. However, Evans took over as president of Ryan Aircraft and James Work was named vice president.

"When D.A.C. bought Mahoney-Ryan," recalled Charles Berkmeyer, former treasurer-auditor," everything was included except a water line to the airport which Phil Ball had the City of St. Louis run from the end of their mains. It was their plan to have the City buy this and use the funds to pay for the liquidation of the Company. The balance, if any, was to go to Mahoney-Ryan stockholders."

Gerry Balzer

B-5 on floats, Missouri River near St. Louis

For this purpose certificates were issued to stockholders which read, in part: "In the event Mahoney-Ryan Aircraft Corporation should hereafter have any assets available for further distribution . . . in addition to the 200,000 shares . . . of Aircraft Development Corporation. . "

It was mid-1929 when Ryan Aircraft Corp. became the fifth builder of Broughams, following Ryan Airlines, Inc., B. F. Mahoney Aircraft Corp., Mahoney Aircraft Corp. and Mahoney-Ryan Aircraft Corp.

Why did Ball, Knight and Bixby sell out to Detroit Aircraft?

"For the quick buck!"

That was the view of Bob Adams, who continued, "I don't think they ever intended to keep it for long and realized the Lindbergh aura was fading fast. They got Evans and some other board members who were forming Detroit Aircraft down to St. Louis and after a prelude of mint juleps ended up with about $2 million in DAC stock".

* * *

Thirty B-5 Broughams were produced by Mahoney-Ryan during the first half of 1929, then on August 27, 1929 workmen rolled out the first product built under their new Ryan Aircraft Corp. name — B-5 Serial No. 218, bearing license NC381K but still with a winged "M" for Mahoney emblem on the rudder.

Art Mankey, chief engineer, had to take the technical data on the B-5 model east to Washington to try and get it approved as quickly as possible after the move to St. Louis so as to keep the factory busy on the newest model. A friend there explained that the Department of Commerce had six months work ahead of Mankey's request for approval of the B-5, but, because of a long standing friendship, the official agreed to accept Mankey's verbal assurance that the supporting technical data was accurate.

Accordingly, Approved Type Certificate No. 142 was granted the B-5 on June 13, 1929 to cover both landplane and seaplane versions. One additional feature incorporated in some B-5 models was the Townend speed-ring cowling circling the engine cylinders as a means of reducing drag and improving performance.

One effort to promote the use of Broughams for business travel was to feature the B-5 as a well-equipped flying office. One such plane was the Brougham purchased by the Prest-O-Lite Storage Battery Company, and described in these words:

Ev Cassagners

Use of planes for business travel was encouraged by Prest-O-Lite's flying office

"A mahogany desk surmounted by a dictaphone occupies one end of the cabin, which is sound-proofed so that the noise of the motor and propeller does not affect the dictaphone. The cabin is fitted with filing equipment, vacuum containers for food and liquids, desk and reading lights for night travel, electric cigar lighter, a folding seat for a second pilot, a cabin heater, non-shatterable glass, and a large baggage compartment."

A considerable number of both the Mahoney-Ryan B-5s and those subsequently produced by the new Ryan Aircraft Division of Detroit Aircraft went to Latin America to open up new air service routes. At least 22 of the 61 B-5s built ended up with Mexican licenses. Seven B-1s and two B-3s had also gone to Mexico.

In August 1929 Pickwick Latin American Airways opened a three-flights-a-week mail, passenger and express service over the 2,700-mile route between Los Angeles, Mexico City and Guatemala City with seven B-5 Broughams. Although a vast improvement over the ox-cart, the first airline schedule had to be flown in a series of daylight flights with overnight stops at Nogales and at Mazatlan on the west coast. Arrival in Mexico City was on the afternoon of the third day.

Records complied by A. Villaseñor of Mexico City show that in addition to Pickwick Latino Americana, the principal Mexican airlines buying Broughams were Corporacion Aeronautica de Transportes of Torreon and Aeronautical del Sur of Villahermosa.

Robertson Aircraft of St. Louis, which had operated contract air mail routes with B-1 Broughams, also purchased the later B-5 model. Bud Gurney, one of their pioneer pilots, reported that "my old log books show I flew the delivery acceptance flights of B-5s with licenses C-735, C-738 and C-739 which were used on the St. Louis to New Orleans run for two years, 1930 and 1931."

Another key market for Broughams was the "Southern National Government" of China in which area nine of the five-place Broughams, mostly B-1s, were sold after embargo on such sales was lifted by the U.S. State Department.

As the result of several earlier long distance flights in B-1 c/n #143 including a circuitous trip of 3,600 miles from Canton to Mukden, Manchuria to Naching, General Chang Wei-Chang became recognized as the 'Lindbergh of the East' and was named Director of Aviation of the War Department.

Five of the Broughams were purchased by the Wuhan Civil Aviation Association for a 700-mile scheduled service between Hankow and Canton on the South China Sea, principally to serve the needs of

Hugh Belknapp

New Mahoney-Ryan factory at Anglum, Missouri, present site of Lambert Field, St. Louis

Rena Vale

Commercial airline service in Mexico was pioneered by Pickwick Latin American Airways in B-5s.

First flight across China was made by General Chang Wei-Chang in a B-1 Brougham

Aero Digest

In Chinese markings, a Brougham is delivered in Hankow by L. E. Gale, left, Mahoney-Ryan distributor

Ryan Aeronautical Library

Ev Cassagners

Hauling one of 11 Ryans through the streets of Shanghai. Planes were to be used on Canton-Hankow route

foreign corporations having large holdings in the southern provinces of China.

Other forms of transportation held little attraction for prospective passengers. Douglas W. Clephane, Mahoney-Ryan publicist, wrote in Airway Age:

"The airline makes the trip in seven hours, while formerly from one to three weeks was required for the very hazardous trip between these points.

"Roads are almost unknown in China. Most of the passengers and freight traffic is by the slow river steamers. The travelers in this country can expect to be held up by bandits, or soldiers of the various revolutionary parties, robbed or required to pay tribute in return for their lives. Even at that they are often shot or dumped into the river in sacks as a form of amusement."

Although Clephane reported that the English, French and German planes previously imported into China were inferior to the Broughams and other American aircraft, firms from these countries were able to offer the Chinese the inducement of better credit arrangements.

Two more Broughams were sold to General Liu of Canton. The name of General Chan Hing Wun also was associated with Broughams in China.

Other Ryans were used for advanced training by the Canton Aviation Bureau. Two of the B-5s exported to China were float equipped for operation from the network of canals and rivers along the Yangtse River and in Kiangsi province.

Sales of Broughams in China were negotiated through the L. E. Gale Company of Hankow, Mahoney-Ryan's distributor.

Gerry Balzer

Larger, more powerful B-7 model was billed as a big sister of the "Spirit of St. Louis"

B-7 and the C-1

In three and a half years, from the first flight of the M-1 to volume production of the B-5 Brougham, each model change showed some useful improvement in utility and performance which served to convince business organizations of the advantages of air travel.

But the B-5 was not to be the ultimate in utility nor the end of the product line. Always the customer sought greater performance. Again the Brougham was scaled up – this time to handle a different radial engine, the powerful 9-cylinder Pratt and Whitney Wasp C-1 of 420 horsepower.

The newest version was the more buxom, extended fuselage B-7 Brougham – advertised as the 'Big Sister' of the *Spirit of St. Louis*. Once again the fin-rudder design was changed. With the much more powerful engine, the B-7 was able to cruise at 140 mph and would presumably attract a new and broader market.

October 26, 1929, Approved Type Certificate No. 262 was issued for both landplane and seaplane versions of the B-7 but only eight planes (Serial Nos. 249-256) were completed at the St. Louis plant of the Ryan Aircraft Division of Detroit Aircraft. Little wonder. The stock market had just crashed, and the great Depression of the early Thirties was settling over the country.

One of the eight B-7s, NC555N, went to Robertson Aircraft for use on their mail/passenger scheduled flights.

When the B-7 made its appearance, Art Mankey again went to Washington, as he had done with the B-5, to expedite approval of the new plane by the Department of Commerce. Later he accompanied Vance Breese on a flight to New York where tests of the seaplane version were conducted at the plant of the builders of the pontoons – Edo Aircraft on Long Island. After some days of tests in that area, approval of the seaplane version was also granted by the Department of Commerce.

One phase of the B-7 Wasp job was, as Art Mankey said, "the only sad experience I ever had in my life."

Mankey had designed the B-7 to use a new type of balance ailerons, known as Freise ailerons. Red Harrigan, the company test pilot who had come from San Diego, flew the plane and was enthusiastic about this design feature. However, the new ailerons were opposed by Bill Schult, one of the key engineers.

Tierney & Killingsworth

B-7 Brougham with speedring engine cowl was used by National Airlines. Pilot is Capt. E. J. Kershaw

Schult was a personal friend of Donald Hall's and perhaps felt that Mankey had not been fair to Hall, whose place he had taken in San Diego as chief engineer.

In any case, says Mankey, "He was undermining me all the time with Phil Ball and his son-in-law, John Nulsen." Without Mankey's approval Schult put another type of ailerons on the B-7. Harrigan didn't like what he saw and after a flight came storming into the drafting room.

"I never saw a madder Irishman in all my life," recalls Mankey. "He said, 'Art, get old man Ball up here and John Nulsen and get Bill Schult, too.'

"I got them there.

"Red said, 'I want you people to know here and now that I want those dammed ailerons Bill designed out of there, and the ailerons that Art designed put back on the B-7.'

"And, by golly, they did!"

In a lighter vein, Mankey recalls how dead set Phil Ball was against smoking.

"John Nulsen was a good egg, so he didn't object when we up in the engineering department smoked. But when his father-in-law showed up it was another story.

"John would come up and say 'the old man is about due and you'd better put those things out.' Ball would show up, sniff around and invariably smell my cigar. He generally made quite a scene which always ended up with a short but positive reprimand on the evils of smoking."

* * *

With the turn of the New Year in 1930, Ryan ads for the first time began to appear in aviation trade journals over the logotype signature of Detroit Aircraft. One of the first planes to be so advertised was yet another version of the family of Ryan cabin monoplanes, the C-1 Foursome.

Instead of a scaled-up version of the latest Brougham, it was scaled down to try and fill the market for a 'Baby Sister' of the *Spirit of St. Louis.* Price was advertised as $9,985 for the four-place C-1 compared with $16,900 for the recently introduced Wasp B-7 model. Powered with the reliable Wright Whirlwind J-6 engine uprated to 240 horsepower, it provided performance comparable to that of the B-7.

The C-1 retained the distinctive Brougham look and incorporated many minor refinements in design and decor, no doubt stimulated by the automotive influence of the new Detroit owners. The brochure copy was also straight out of Detroit.

"The new Foursome," it read, "is designed particularly for the business executive, private owner or sportsman who wants the comfort and reliability found in the limousine, private yacht or pullman car, combined with the high speed and freedom from transportation schedules that only a privately owned aircraft can offer." But in the gloomy economic climate of early 1930 it did not sell.

The C-1 was designed under Art Mankey's direction and built by Detroit Aircraft in St. Louis, receiving Approved Type Certificate #346 on August 11, 1930. Meanwhile Mankey had moved to Detroit as assistant to Dayton Brown, chief engineer of Detroit Aircraft, and Bob Adams had gone there to head up advertising and promotion.

The Foursome models were given the 400 series of company serial numbers but unfortunately the count of planes built never got above three so that c/n 402 was the last Brougham built.

Although the record of the three C-1s is obscure, one which had been purchased by the Horlick Malted Milk Co. was acquired in 1938 by Alexander Loeb and Richard Decker, fitted with extra tanks and christened *Shalom* for long distance flight.

Trying to take off in July 1939 from Floyd Bennett Field, Brooklyn, with a

Ryan Aeronautical Library

Baby sister of other Broughams was the C-1 Foursome developed under Detroit Aircraft banner

Warren D. Shipp

C-1 christened "Shalom" takes off from Nova Scotia on ill-fated flight

heavy gas load, Loeb and his copilot burst a tire on the C-1 and were grounded by the CAA for allegedly attempting a flight to Palestine. The *Shalom* next showed up in Nova Scotia. There Loeb and Decker took off from the beach at St. Peter's on August 11 and were never seen again, having no doubt been lost at sea.

In the last week of July 1930 the relatively new 'Mahoney-Ryan' plant at St. Louis was closed by James Work, general manager of Detroit Aircraft and what little was left of Ryan Brougham production was moved to Detroit Aircraft's plant in the motor city. There one last effort was the installation of the Packard Diesel aircraft engine in the second C-1 built, which was then redesignated a C-2 model.

The very last C-1 Foursome built was reportedly purchased by test pilot Vance Breese for his own use.

* * *

"It was useless to continue making planes under present conditions," explained John Nulsen, manager of the St. Louis plant. "Sales have been disappointing since last fall, and it was futile to continue." Nulsen and most of his associates terminated their connection with what was left of the original Ryan organization in mid-August. Hugh Belknapp, however, stayed on for a while to handle orders on spare parts for the several hundred planes out in service.

Although he was a director of Detroit Aircraft, Phil DeC. Ball and his associates were disappointed to see airplane production moved out of St. Louis. Civic pride was somewhat restored when Ball repurchased the Mahoney-Ryan plant at Lambert Field from Detroit Aircraft, and even more so somewhat later when he bought the bankrupt Mono Aircraft Corp., headed by Don Luscombe, and the Lambert Engine Company and moved both from Moline, Illinois. Still later J. D. Wooster Lambert ,an original Lindbergh backer, became president of the reorganized Monocoupe Corp., and John Nulsen moved in as that company's general manager.

No doubt out of friendship for the backers of his New York-Paris flight, Col. Charles A. Lindbergh ordered a Monocoupe for his personal use since he was no longer flying the Special B-1X Brougham which had been a gift to him in April 1928.

One little-known airplane of the period which, for a time, bore the Ryan name was the Speedster biplane. It was not truly a Ryan product.

Parks Air College, one of the 12 firms in the Detroit Aircraft group, had developed the Parks model P-2A Speedster as a classroom project. It received its approved type certificate in November 1929. When aircraft production was moved out of St. Louis by Detroit Aircraft Corp., the former Ryan and Parks operations were consolidated in the Detroit, Michigan plant and the name Ryan was attached to the Parks-developed biplane.

* * *

And what was the fate of Detroit Aircraft?

The ambitious effort to make Detroit the aircraft capital of the world was falling apart.

Whistling in the dark, John Nulsen had announced soon after the 1929 stock market crash that Ryan Aircraft planes would soon be on sale and exhibition in all American cities of over 10,000 population and that, for the first time since Lindbergh's flight, the company was able to make reasonably prompt delivery on planes. But as the Depression deepened, unsold airplanes began stacking up in plants around the country until some 1,100 of all makes were in need of buyers.

Only Lockheed among the Detroit Aircraft subsidiaries kept its head above water — for a while. During 1930 Lockheed continued assembling planes in California, with metal fuselages for several models being shipped to Burbank from the Detroit plant. By 1931 all other aircraft production had come to a halt.

One project did show some promise. It was the Detroit/Lockheed XP-900 ex-

Lindbergh, left, ordered a Monocoupe for his own use. John Moliter, later of TWA, right

Hugh Belknapp

Walter J. Kohler, Governor of Wisconsin, an ardent aviation backer, pictured with Wasp-powered B-7 Brougham

Missouri Historical Society

Hugh Belknapp

Detroit Aircraft executives with XP-900. Left to right, Peter R. Beasley, president; William Pritchard, factory manager; Art Mankey, chief engineer. In cockpits, Robert Woods, project engineer, and Vance Breese, front, test pilot

Hugh Belknapp

Detroit/Lockheed XP-900 went on to become Consolidated PB-2A pursuit

perimental pursuit plane built in 1931 under contract to the Army Air Crops. The all-metal fuselage and empennage was built in Detroit. The conventional Lockheed wooden wing had been built in Burbank and shipped by rail to Detroit.

The plane, powered by a 12-cylinder V-type Curtiss Conqueror liquid-cooled engine, was assembled at Groose Ile Airport, which was also owned by Detroit Aircraft. It was one of the first aircraft with retractable landing gear. Project engineer was Bob Woods, later chief engineer at Bell Aircraft.

By a strange coincidence, Vance Breese who had been the demonstration pilot on the first Ryan plane, Claude Ryan's M-1, was also the test pilot on the last of the Ryan-related companies — Detroit Aircraft's — final products. Only the one XP-900 was built and it was destroyed when an Army pilot bailed out on a test flight.

"The XP-900 could possibly have been the salvation of the company," was the opinion expressed by Hugh Belknapp. "As I recall the company had a contract with the U. S. Army Air Crops for about 20 planes, and the project appeared very promising. After the crash of the first plane, difficult financial conditions forced the company further toward receivership from which there was no recovery. Thus the end of the plane and eventually of Detroit Aircraft Corporation."

On the West Coast Lockheed did manage to struggle along on new models which found a few customers among airlines, the military, and flying greats like Lindbergh and Jimmie Doolittle. Then even this trickle of orders stopped.

By April 1932 employment at Lockheed was down to four, and business prospects had vanished. Undaunted, a group headed by Boston investment banker Robert E. Gross, who had spent several years each with Stearman Aircraft Company and with Viking Flying Boat Company, scraped together $40,000 and bought Lockheed's assets. He managed to turn the company around with development of a new all-metal airliner for the expanding transport market.

Claude Ryan, too, gave some thought to buying the Ryan Aircraft Division out of the Detroit Aircraft receivership but rightly decided their product line was getting obsolete. By the following year Claude Ryan was busy developing the modern new Ryan S-T metal-fuselaged low-wing sport trainer which was destined to become one of the great classic airplanes of all time.

The liquidation — the last days — of Detroit Aircraft were presided over by the court-appointed receiver, Detroit Trust Company, whose trust officer in charge was a huge Scotsman named Archie Campbell. Charles Berkmeyer, the Mahoney-Ryan treasurer, originally from St. Louis, tried to help Campbell untangle the records and few remaining assets of DAC and its many subsidiaries. In turn Berkmeyer was assisted by Hugh Belknapp, also from St. Louis, in taking inventory of and assisting in disposing of some of the physical assets.

While working in St. Louis, Belknapp lived in the same old farmhouse on the edge of Lambert Field which Lindbergh had occupied while flying the airmail as a pilot for Robertson Aircraft. "It was," says Belknapp, "rather primitive living but we enjoyed it as it was within easy walking distance to the Ryan plant."

Thanks to Belknapp, although he disposed of thousands of pictures and countless records, he retained a good many items of historical value which have filled in gaps in the information previously available about the Ryan Aircraft-Detroit Aircraft era.

* * *

What went wrong with the Brougham and the six companies which manufactured the most popular plane of its day?

Surely no company had a brighter prospect following Lindbergh's flight in the *Spirit of St. Louis,* than did Ryan Airlines, Inc.

Was it possible for any company — despite its great potential — to have weathered the economic storm of the Depression?

Forty years after Detroit Aircraft went down in receivership we put the question one day to Bob Adams, whose subsequent career gave him some mature perspective on the matter.

After Detroit folded, Adams did some flying around New York City then went to Wright Aeronautical Company as advertising manager, remaining there for four years until opening his own agency, Burke Dowling Adams (his full name, not those of three partners). His initial account was Wright Aeronautical.

C. E. Woolman of Delta Air Lines liked Adams' ads for Wright so much he gave Adams' agency their account in 1945. In 1951 Adams moved his agency to Atlanta, headquarters city of Delta.

With that kind of aviation/advertising background to his credit, Bob Adams in 1972 looked back on the Brougham days —

"I've often though about what the company needed after the move to St. Louis. For one thing, we did not have a leader like Claude Ryan or Walter Beech or Clyde Cessna who could fly and contribute to the design.

"The engineering department consisted of Art Mankey and two assistants, the latter little more than draftsmen. The policy was to make minor improvements to the basic Brougham but no great changes — dual controls, six seats and the Townend speed ring being the major airframe changes. I jumped into the design of the exterior before the Detroit Air Show in 1929 and got them to put more automotive styling and finish on the cabin area.

"Of course, they had Don Hall's X-1 design from San Diego but the *Doodle Bug* was not a thing for student pilots. We should have had something like the Monocoupe or the Ercoupe with a broader market to keep the plant open and meanwhile work on a larger, all-metal design for early airline use. Carl Gray, President of the Union Pacific Railroad, told me early in 1929, "Come back and see me when you've got a plane that will carry 20 passengers.'

"The main trouble was there was little leadership and direction and a lack of engineering clairvoyance, coupled with management by promoters who were looking for a quick profit. I guess the fact that the Depression was so widespread would have been hard to surmount but some aircraft companies did manage it, notably the original Ryan west coast operation."

In later years, Lindbergh, too, had some observations about the Brougham. "Donald Hall," he wrote, "felt strongly that the Ryan Brougham was not adequately designed. He felt it was a mistake for the company to go into production with this design; but it was a reasonably good plane on the standards of the time. Mahoney felt there wasn't either money or time to go through the procedure of redesigning."

During his 1927 tour of the United States in the *Spirit* — as a promotion for aviation sponsored by the Guggenheim Fund — Lindbergh had been giving some thought to future aircraft design.

Thus, from Little Rock, Arkansas, Lindbergh wrote to Donald Hall on October 2:

"I have been thinking a little about the best type of plane to be built around a Whirlwind motor.

"I thoroughly agree with Mahoney's idea of keeping the present carrying capacity and increasing speed rather than increasing the payload at the present speed.

"I am not all sure that the parasol type of monoplane is best. Visibility could be increased by having the fuselage above the wing and a folding landing gear could be installed even more easily.

"The possibility of a pilot landing with the gear folded could be minimized by installing a signal on the throttle. A ship is never landed with the throttle open nor flown with it closed. Consequently it could be arranged for a conspicuous signal to appear if the throttle was closed with the gear folded or, in fact, it might be arranged so that the throttle could not be closed with the gear folded. In case of emergency the switch could be set.

"I would like to get your idea on this."

But nothing came of Lindbergh's suggestion.

Last gasps of dying Detroit Aircraft and its Ryan Aircraft Division came at this Detroit plant

Hugh Belknapp

Richard Stauss

"Spirit II" takes shape in Tallmantz shop under watchful eyes of Claude Ryan and John van der Linde of original Ryan Airlines and Frank Tallman

Replicas of the "Spirit"

Forty years after Lindbergh flew the Atlantic in the most famous Ryan Brougham of them all, the 'assembly line' went back in production for one more plane — an exact duplicate of the *Spirit*.

It was 1967 and the United States Government was planning its participation in the Paris Air Show around an 'In the Spirit of Lindbergh' theme.

What would be more appropriate on the anniversary date than to recreate Lindbergh's landing at Le Bourget Airport on the night of May 21, 1927. Weren't there one or two Broughams rebuilt as replicas — close imitations — of the *Spirit* for the Jimmy Stewart movie still around?

Those replicas for the 1956 film had been provided by famed movie pilots Paul Mantz and Frank Tallman. The movie itself had its premier showing in 1957 on the 30th anniversary of the historic flight. In the interval Mantz had been killed in a movie accident but Tallman was still operating Tallmantz Aviation.

Yes, Tallman told the Department of Commerce coordinator for the Paris Air Show, he would see that one of the movie replicas was available to again fly around the Eiffel Tower and land at Le Bourget. In fact, he'd fly the plane himself. It was the least Tallmantz Aviation could do in helping the government recognize the 40th anniversary of Lindbergh's flight.

But it didn't quite work out that way. Apparently the replica Tallman planned to use had been sold or mortgaged to Nebraska grain and auto dealers. Frank assumed the owners would go along with the government's request and donate use of the plane, but the owners held out for an unrealistic rental fee, and the government did not have the funds in its Air Show budget.

Tallman's reaction was a "to hell with it" attitude — "so I'll build a new, exact duplicate myself." If Ryan Airlines could build the original in 60 days, Tallmantz could build the duplicate in 90 days.

Fortunately the San Diego AeroSpace Museum already had a similar plan, "Project We," in the concept stage under the leadership of Dan Burnett and Ed Morrow, two retired Ryan employees, both of whom had worked on Lindbergh's plane. Their working drawings were nearly complete; they made them available to Tallman.

Work got under way at the Tallmantz shops at Orange County Airport in Southern California in January. Tallman knew of a Ryan M-1 fuselage which was at Lancaster airport, bought it and used it in helping recreate the *Spirit of St. Louis II*.

A team from Ryan Aeronautical Company headed by George Becker took over coordination with the Department of Commerce and Paris Air Show people.

In a letter to General Charles A. Lindbergh in mid-summer 1967, T. Claude Ryan picks up the story:

"Frank Tallman's small crew of old-time skilled mechanics went to work. He put in his own money and they did the job in a remarkably short time, including evenings and weekends. Hawley Bowlus, Don Hall, Dan Burnett, John Van der Linde and Walt Locke were interested 'consultants.'

Richard Stauss

Forty years after Lindbergh flight, "Spirit II" and Eiffel Tower recreate historic event at Paris Air Show, 1967

"On April 24, 1967 I watched Frank test hop the *Spirit II* from a grassy strip between runways at Orange County Airport. It was quite a thrill and surely brought back a lot of memories. (The original *Spirit* made its first flight April 28, 1927—Ed.)

"Frank later flew the duplicate to North Island, San Diego, and made a symbolic flight the next day, May 10, re-enacting your takeoff from there for St. Louis exactly 40 years before. This was of special interest to a dozen or so of the men and women workers present who had helped build your plane.

"The plane landed at Lindbergh Field where it was disassembled and loaded in an Air Force C-141 transport. Frank and one of his mechanics went along, following your original route through St. Louis.

"The plane was assembled at the airport outside Paris and on May 21st duplicated your landing at Le Bourget for the thousands of happy Parisians again gathered there. Later it was mounted on a pylon at the entrance to the U. S. Pavilion.

"A number of San Diegans became so enthused about having seen a genuinely accurate duplicate of this most famous plane that the "Lone Eagle Corps" bought it for $40,000 for permanent display when the C-141 returned from Paris last week. It is now the proudest display in the San Diego AeroSpace Museum. Right alongside is its predecessor, the sole remaining Ryan M-1 monoplane."

Richard Stauss

Spirit replica becomes symbol of U.S. theme "In the Spirit of Lindbergh"

Unlike the movie replicas, the *Spirit II* was as near to being an exact duplicate of the original as was humanly possible — recognizing that in the 40 intervening years certain equipment items were no longer available.

Authenticity was especially important in the cockpit area. Lindbergh was seated well aft because he wanted the gas tanks between himself and the engine in case of an accident during takeoff with the heavy fuel load. For this reason there was no windshield and hence vision only out the sides of the fuselage. The small periscope he had for forward vision was seldom used by Lindbergh.

Tallman saw that these features were exactly duplicated. Of the three movie replicas some had the pilot seated forward in the normal cabin plane arrangement, usually with a partial forward windshield and cockpit windows on the side opposite the camera. In another important respect the movie replicas were not always accurate. Some retained the 42-foot wing of the standard commercial Brougham. Tallman's *Spirit II* had the same 46-foot extended wing as the original.

When Warner Brothers Studio undertook production of the motion picture based on Lindbergh's Pulitzer Prize-winning book they found in Jimmy Stewart, himself a World War II pilot, the ideal actor to play the part. But with the original *Spirit* in the Smithsonian where would they get the second half of 'We' to duplicate the famous Ryan monoplane's role?

Paul Mantz, best known of movie pilots, was just the man to find, build and fly two of the planes they needed. Jimmy Stewart had some ideas, and he too was part owner of yet another plane rebuilt for the movie. Here is the story of the movie reproductions.

With three replicas available, the film producers would be able to speed up production by making close-ups in the studio at Burbank, have one crew shooting in Paris, and another making over-water scenes off Half Moon Bay near San Francisco.

It was simpler, the studio had said, to change Jimmy Stewart to Lindbergh than it was to duplicate the *Spirit*. Both men have the same lean appearance; Stewart merely switched the part in his hair from right to left and lightened the color slightly to become Lindbergh.

The problem of finding the 1927-28 vintage Ryan B-1 Broughams was solved by a check of the latest aircraft listings.

In a scene markedly similar to original mating of wing and fuselage (see page 24), Tallmantz workmen assemble replica of Lindbergh's plane

Richard Stauss

Ryan Aeronautical Library

Warner Brothers

Both men, Lindbergh, left and Jimmy Stewart, right . . . have the same lean appearance

Air Force C-141 gobbled up Spirit II and flew it to 1967 Paris Air Show

Richard Stauss

Paul Mantz, beneath propeller, built two of the three "movie" replicas used in 1957 motion picture epic

Ryan Aeronautical Library

Both the two craft purchased for Mantz by Stan Reaver, his chief pilot, were in flyable condition.

The first of these, Serial No. 159, was located in February 1955 in Moscow, Idaho, Licensed as N7212 (NC7212 until 1948), it had last been used to ferry hunters into nearby deer country. Reaver and mechanic Gene Wells flew it south to Orange County Airport where it was rebuilt.

The second Mantz Ryan was c/n 153 (NC7206) purchased from a mechanic working for McDonnell Aircraft in St. Louis. His plane, however, was located in Caresco, Michigan. The ferry flight from there to Southern California took Reaver and Wells 23 hours and 20 minutes. One old entry in the log book showed "128 passengers carried in three hours of flying in one day" near Danville, Illinois.

Meanwhile Mantz had scrounged up six Wright J-5 220 hp Whirlwind engines and was able to get four completely rebuilt power-plants out of them.

The third rebuilt B-1 was Serial No. 156 (NC7209). Its owners were movie star Jimmy Stewart and his friend, Hollywood insurance executive Joe DeBona. W. H. (Hank) Coffin had found it literally in shreads, in Greeley, Colorado, where it had long been staked down in the open. It had not been flown in the three years since it was used to ferry fishermen into the high meadow country of the Rockies.

Don Downie

Parke-Bernet art galleries auctioned off this replica to David Jameson who displays it at EAA Museum, Oshkosh, Wisconsin

Coffin and partner Paul Franklin purchased it for $1,000 and trucked it back to Southern California. After it was rebuilt and flying one can imagine the reaction of the airline pilot coming into Los Angeles International Airport when the tower radioed, "DC-6 on final approach — just follow the *Spirit of St. Louis* ahead of you."

And what became of the three movie *Spirits?*

Jimmy Stewart made a gift of c/n 156 to the famous Henry Ford Museum and Greenfield Village at Dearborn, Michigan.

The second Mantz replica, c/n 153, was purchased in 1963 for $11,000 by the Missouri Historical Society and currently is stored in a McDonnell-Douglas hangar at Lambert Field, St. Louis. Two of the donors, J. D. Wooster Lambert and Harold M. Bixby of New York, were among the original backers of Lindbergh's flight.

The first Mantz replica, c/n 159, was placed on the auction block by Tallmantz Aviation in a highly publicized sell-off of vintage aircraft in May 1968 conducted by the famous New York art auctioneers, the Parke-Bernet Galleries. Purchaser, for $17,500, was David Jameson of Oshkosh, Wisconsin. It is now on loan for display at the Experimental Aircraft Association Museum in Oshkosh.

Joe DeBona

McDonnell-Douglas

Jimmy Stewart's movie replica is displayed at Henry Ford Museum, Dearborn, Michigan. Another replica, lower, is owned by Missouri Historical Society, St. Louis

Bill Monday

After landing at 9400 feet elevation in Yellowstone, saddle horses pull the Brougham to hill top for takeoff with sick passenger

Pilots and Planes

Although Brougham production came to a screeching halt within a scant five years after development of the M-1/Brougham concept, the dependable cabin monoplanes provided excellent service for many years after they were out of production.

Bill Monday's B-5 was an excellent example.

By the mid-thirties Monday had gone into the flying service business for himself in Cody, Wyoming, the eastern entrance to the high Yellowstone Park country.

Broughams were the ideal plane for heavy duty flying and Monday found just what he required in a used Ryan B-5 Brougham powered with the 300 hp Wright J-6 engine. NC-728M, serial No. 228, proved to be just what Bill needed. It was in perfect condition although second hand, and was purchased in May 1936 for a rock-bottom $1,000 from its prior owner, R. A. Stilitz of Philadelphia.

One of Monday's good friends was Charles J. Belden, owner of the 200,000-acre Pitchfork Ranch at Meeteetse, Wyoming, 35 miles south of Cody. Belden was not only the operator of one of the largest ranches in the west but also a talented photographer and early wildlife conservationist.

Years earlier the ranch had been stocked with a herd of 25 prong-horned antelope but with the passing of time had grown to over 4,000, making them a serious threat to the corps intended for winter feed of the ranch's cattle and sheep. Each fall Belden opened the ranch to hunters as the herd had to be depleted to managable size.

But in August 1936 Belden had a better idea for handling at least some of the fawns. He had trained a dog to capture fawns which had become separated from their mothers. The dog would chase the stray fawns, bowl them over and pin them to the ground, but never bite them. In this way 23 of the two-month old 20-pound fawns were caught.

Loading them aboard the B-5 Brougham, from which the rear seats had been removed, Monday and Belden took off on a mission of supplying eastern zoos with prong-horned antelope. After a 12-hour flight from Cody to Lambert Field, St. Louis, three fawns encased in individual burlap sacks from which their heads and necks protruded, were delivered to the St. Louis Zoo in Forest Park. Earlier two had been given the St. Joseph, Missouri zoo during a refueling stop there.

Whitney Galley of Western Art, Cody, Wyo.

Charles J. Belden of Pitchfork Ranch with 23 antelope to be delivered by Brougham to zoos

Mountain flying along the Salmon River of Idaho

Charles J. Belden

Similar deliveries were made to zoos in Chicago, Washington, D.C., Philadelphia, and Newark, the latter for the Bronx Zoo. Finally Belden and Monday landed at Lakehurst, N. J. where the famous trans-Atlantic Zeppelin *Hindenberg* waited to fly two of the fawns overseas to the Berlin Zoo — just five days out of Cody.

Later Monday and Belden teamed together on many flights including taking fishing parties to the Hood Ranch, 100 miles northeast of Boise, Idaho, on the middle fork of the Salmon River. Pictures Belden took there in the summer of 1937 are among the best of early recreational uses of the airplane.

Monday's B-5 was one of the most picturesque of the Ryan Broughams. By then he had equipped the B-5 with a Townend speedring and on it had painted brands of the different cattle ranches in the area and the rudder bore the official 'bucking horse' design for Wyoming state automobile license plates, used on the plane by special permission.

The unique capabilities of the Brougham made it ideal for emergency and rescue flights because of the long distances between the small towns of Wyoming, Montana and Idaho and the remoteness of the few settlements in the higher mountains.

Fishing and hunting trips into otherwise inaccessible country with local guides and eastern sportsmen were also a source of regular business for Monday and other pioneer pilots.

Lester Jordan and Guy Murphy of Kansas City found out the hard way how useful Bill Monday and his B-5 could be. While hunting on the upper Yellowstone River, Jordan became ill and in several days was down with a serious case of pneumonia. Getting him out by horseback would be a long, hard and dangerous task.

Murphy, who knew something of flying, felt that Jordan could be taken by airplane from a meadow near their camp. Riding all night, their guide was sent out by horseback to the TE ranch where a message could be phoned to Monday in Cody.

After taking off from Cody in the Brougham, Monday first stopped at the TE ranch strip to pick up Evan Holman, the guide, then landed at Bridger Lake from which point Holman was to go on foot up Butte Creek to the campsite. As reported by the Cody Enterprise of October 6, 1937 —

"After some delay Monday decided to take a look over the field and approached the camp on Butte Creek to find that Murphy and the owner of the TE ranch had marked out a field for him in a small revine, and he maneuvered the Ryan to the emergency field. (A steep hillside at 9400 feet altitude. — Ed.)

With the aid of saddle horses, the machine was then pulled to a vantage point for a take-off, and after taking on the sick man, delivered him to Cody in short order.

"Those knowing the Butte Creek section are still scratching their heads and trying to figure out how Monday got his plane in and out of that area, but the stunt was accomplished even though Monday wants little of it as a steady diet.

"There is the fairly good field which has been laid out near Bridger Lake, but landing on any of the smaller creeks at the altitude, which in most cases is in excess of 8500 feet, is quite dangerous. Runways naturally would be very short, and timber growths make getting down to the few meadows unusually dangerous. Only in emergencies are planes permitted to land in the Yellowstone."

Once Monday got his sick passenger out of the mountains and back to Cody he fired up his Ford trimotor and flew members of the party back to Kansas City.

Bill Monday

An old timer encounters modern cowpilots

On another occasion, typical of mercy flights, Bill Monday saved the life of John Henderson, a middle-aged sheepherder. Henderson lost his sheep during a blizzard and started for the Heald Ranch to get help. In the blinding snow he became lost and wandered without food or shelter for three days before finally staggering into the ranch when the storm abated .

Although suffering from exposure and from frozen hands and feet Henderson could not be taken into Cody because all roads were closed. Monday and his B-5 were called for and Bill landed in 18 inches of snow near the ranch. As soon as an area could be cleared for takeoff, Monday left for Cody where the highway department kept the runway cleared until he could land the Brougham and get his patient to the hospital for emergency treatment.

Wild horses on the high plateau country were another problem Bill Monday tackled with the B-5. For years a large band of wild horses had run at large in the rugged McCullough peaks country east of Cody. Many riders had attempted to drive the mustangs out but the rough, jagged area balked top cowboys of the Buffalo Bill country in their efforts to round up the herd. Bill Monday was hired to try his hand.

Flying the Brougham at low altitude, Monday flew over the rugged area, located the 'broom-tails' and chased the mustangs toward open, flat country where horsemen succeeded in corraling nearly a hundred head including a handsome stallion which had been branded eight years earlier but not seen since.

One Labor Day, Monday was carrying passengers in the Brougham at a celebration at Meeteetse. A local cowboy, Jerry O'Leary, who "possibly had a drink or two of something stronger than Greybull River water," decided he could put most bronc riding champions to shame.

Why not ride the Brougham cowboy style? In his state, O'Leary thought it would be a cinch. Sure enough, when Monday taxied out in the Brougham with a full load of sightseeing passengers, O'Leary jumped astride the fuselage just forward of the fin.

As soon as the Brougham was airborne Monday knew something was wrong because he couldn't keep the tail up. Quickly he urged the rear passengers to move forward in the cabin while he put the plane in a slow, gentle circle back to the airstrip where he landed safely. Only then did he discover the cowboy stowaway riding the fuselage.

As the Cody Enterprise reported, "The event proved a real thrill for those who witnessed the incident, and there is one cowboy over the Greybull who will keep to the broncs, rough as some of them may be, in the future."

Bill Monday

Wildlife surveys for the government were made by Brougham flown out of Cody, Wyoming

Charles J. Belden

Bill Monday's "bucking horse" Brougham and another from Boise bring sportsmen to Hood Ranch, Idaho

When two U.S. Forest Service rangers making a timber survey in mid-November 1971 in a remote snow-covered area at 5500 feet elevation near Lookout Pass in Idaho's Bitterroot Mountains came upon wreckage of a 1928 aircraft they helped unravel a 27-year-old mystery.

Returning to Kellogg, Idaho, 25 miles distant, they brought out the identification number and instruments of the substantially damaged plane and reported the crash to the Federal Aviation Administration at Spokane, Washington. The body of the pilot of the wrecked plane was not located and it was obvious further investigation would have to wait for the Spring 1972 thaw.

International Airways operated this B-1 in North Dakota before its 27-year disappearance. Note Indian good-luck swastika.

Cecil Shupe

Then started the research which some days later identified the plane as a Ryan B-1 Brougham, license NC-6955, registered to a Leo Watson of Eau Claire, Wisconsin. Further checking disclosed that Watson had sold the plane for $1,700 in 1944 to Wilbert George Carroll, 30, and Sheldon G. Knudson, 28, both of Seattle. The new owners had flipped a coin and it was decided Carroll would fly west while Knudson would follow-by car.

Carroll left Eau Claire, Wisconsin, in the Ryan Brougham in mid-December, 1944 for Seattle. He checked with the earthbound Knudson during stops at Miles City and Billings, Montana. Somehow, at Billings, the plane's propeller was bent and Carroll was advised not to fly on until repairs had been made. But it was December 20, 1944 — just five days before Christmas — and Carroll was anxious to get home to Seattle.

Despite a storm coming up, Carroll fueled the plane after a stop at Missoula, Montana, loaded the cabin with Christmas gifts and headed directly into the approaching snowstorm. The plane was last seen flying so low over Superior, Montana, its license number could be read; then nothing further was heard of NC-6955.

When Carroll failed to show up in Seattle Knudson notified the Coast Guard and Air-Sea Rescue people and a search was started, but it was 27 years before the plane was found.

Knudson was still an active pilot in 1972, flying for the Forest Service out of Ramona California. Working with the FAA and the family of the lost pilot, the San Diego Aerospace Museum organized a trip into the rugged Idaho mountains in Summer of 1972 to bring the plane out. It had been hoped to restore the plane but, not unexpectedly, there was not enough left, with the exception of the engine, to make it worthwhile.

NC-6955 had come off the B. F. Mahoney production line in San Diego on July 31, 1928 as Serial No. 140. It was sold to International Airways, Inc., of Minot, North Dakota, of which Richard C. Coughlin was president. Coughlin was also head of Westland Oil Company which owned and operated M-1 No. 23, which was traded-in on the new Brougham.

The new plane had been purchased by Minot business men to start a scheduled air service between Minot and Bismark, North Dakota. It was also used for charters, primarily to Winnipeg and Minneapolis. Subsequently the plane changed hands a number of times before its mysterious disappearance in 1944 was solved after 27 years.

RYAN BROUGHAM

(First Ryan Brougham)

B-1

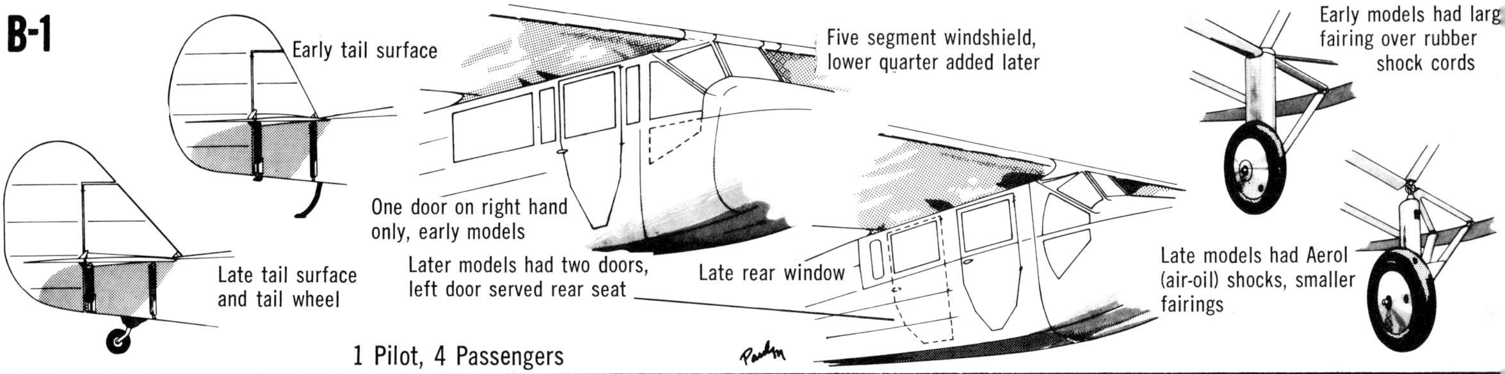

B-3

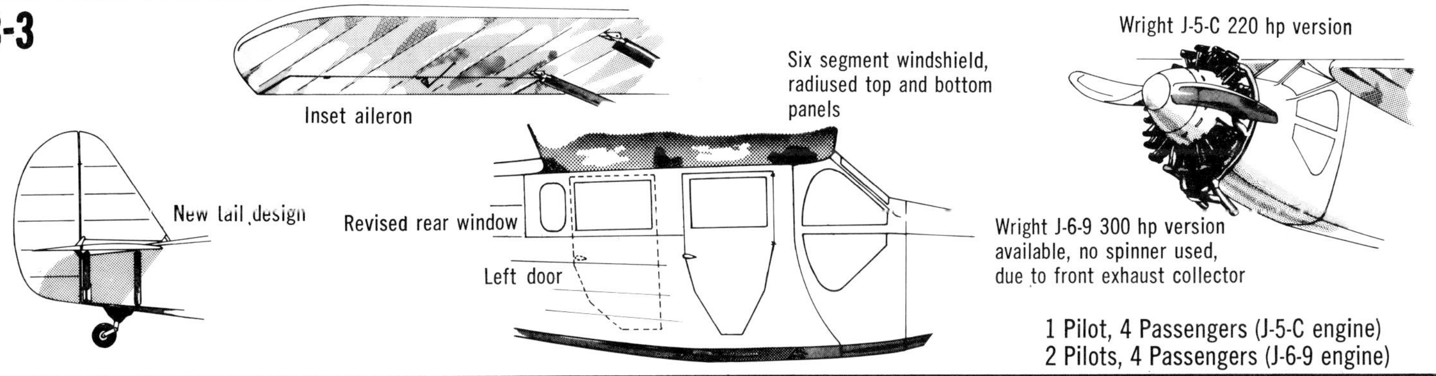

B-5

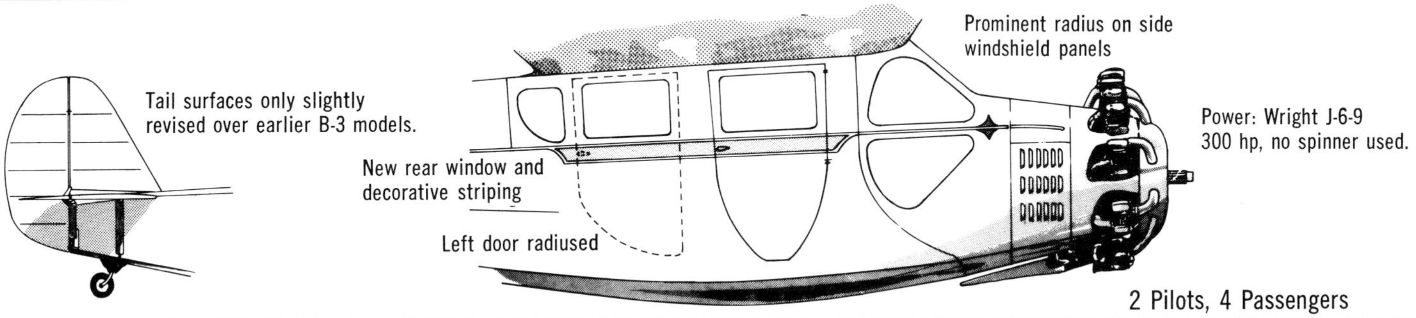

B-7

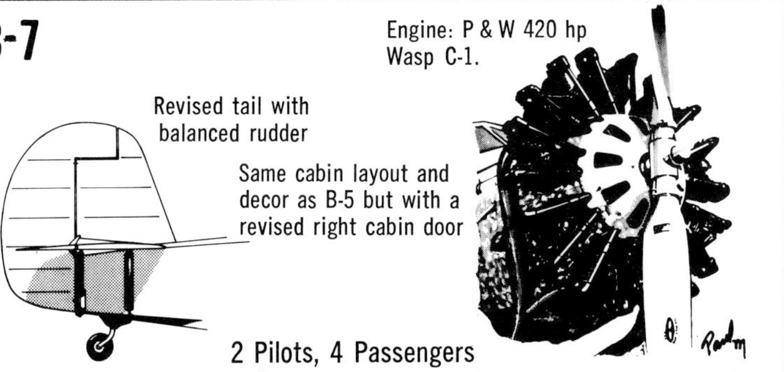

C-1

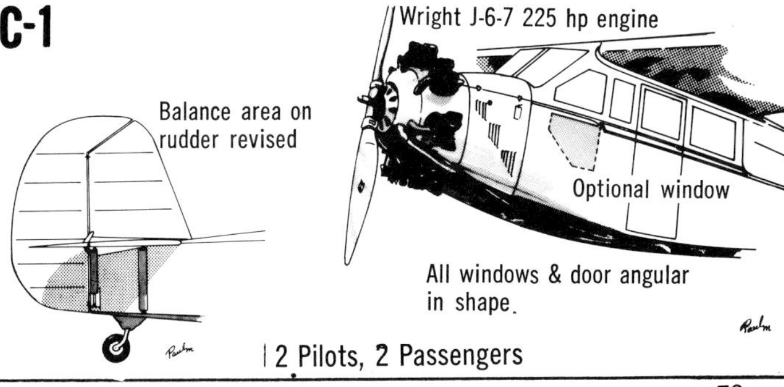

Postscript

If, in the preceding chapters, the author was to divert from the basic history of the Ryan Broughams and their builders to spin other yarns it would be so confusing as to obscure the main story. Yet there are many fascinating related sketches and biographical briefs which make interesting additions to the main history.

First flight of Ryan Mechanics' CM-1 "Lone Eagle" at Vail Field, Los Angeles

Ryan Aeronautical Library

RYAN MECHANICS MONOPLANE CO.

The story of the three founders of the Ryan Mechanics Monoplane Co., and that of pioneer aeronautical engineer William J. Waterhouse, add interesting, but poignant, footnotes to early aviation history.

Few people understood the difficult problems pioneer builders had to face in the post-Lindbergh era of explosive aviation expansion followed by the stock market crash of 1929 and the subsequent Depression of the thirties.

Feeling that their contributions in helping build Lindbergh's *Spirit of St. Louis* Ryan monoplane had perhaps gone unrecognized, and disturbed that B. F. Mahoney was considering moving the original Ryan Airlines company from San Diego to St. Louis, three employees decided to strike out on their own.

Lieutenant Oliver R. McNeel, an early employee of T. Claude Ryan's soon after he founded Ryan Flying Co. in 1922, had been final assembly foreman during building the *Spirit*. B. A. Seitz, a pioneer of the Yukon, aviation buff and Los Angeles investor, had told McNeel he would be interested in putting up $10,000 to finance a new company to build airplanes in that city. Associated with McNeel in the new venture were fellow Ryan Airlines employees Fred C. Ayars, head of the covering and finishing department, and Henry W. Hunold, welding foreman.

Identifying themselves as "the mechanics who built Lindbergh's *Spirit of St. Louis*," McNeel, Ayars and Hunold founded the Ryan Mechanics Monoplane Co. McNeel was president; Hunold, vice president and Ayars, secretary-treasurer. Unfortunately none had any real business experience, the same situation in which B. F. Mahoney found himself after he bought out Claude Ryan and was suddenly swamped with urgent business problems resulting from the great demand for Ryan Brougham monoplanes following Lindbergh's dramatic New York - Paris flight in May, 1927.

Late in 1927 the three San Diegans moved to the Warren School of Aeronautics in Los Angeles which then needed capable men to serve in the dual capacity of instructors for the school and craftsmen able to build prototype aircraft from engineering drawings.

In those days, training and manufacturing were often complementary activities in budding aviation enterprises just as they had been at Ryan.

Next door on Slauson Boulevard was the Western College of Aeronautics whose dean was William J. Waterhouse, an engineer who had done work for Douglas, Ryan and other Southern California aircraft builders. His capability, particularly as a stress analyst, was widely respected.

After Lindbergh's flight, the country had broken out in a rash of actual and attempted long distance flights. McNeel had been bitten by the bug, too. From

Missouri Historical Society

Lieut. O. R. McNeel and the modified "Miss Southern California" he flew in New York-Spokane Air Race

the Warren School, McNeel and his associates acquired a Ryan M-1 thought to have originally been used by Pacific Air Transport on its Los Angeles - Seattle air mail run. Or perhaps it was an unsold Waterhouse Cruzair.

In any case when it came out of the Warren School shops as modified for flights of extended range, the vertical fin carried the identification "Waterhouse Cruzair — Rebuilt by Ryan Mechanics Monoplane Co."

It was christened *Miss Southern California* and bore the Warren School insignia in addition to the huge "Ryan Mechanics Monoplane" sign which covered the entire side of the fuselage.

Forty-two years later McNeel wrote that the plane was "an M-1 used by the Warren School for training in airplane construction. The Ryan Mechanics bought the plane from the school and remodeled it to enter the air race from New York to Spokane, Washington. Bill Waterhouse was our engineer.

"We installed a Wright Whirlwind engine, changed the rudder, installed an additional tank where the front seat had been, remodeled the engine cowl and beefed up the shocks.

"After test flying it from the Western Avenue airport in Los Angeles, I flew it to Rockwell Field, San Diego, where we obtained permission through my old commanding officer, Major H. H. "Hap" Arnold, to do additional tests and take off for the east.

"When all the tests were completed, I flew it to New York, leaving San Diego at 5:05 p.m. on September 17, 1927, stopping en route at St. Louis for oil and fuel — precisely as Lindbergh had done four months earlier.

"I took off from New York on the 19th for the nonstop flight to Spokane. Everything was going fine until I reached Rapid City, South Dakota, where the cam ring went haywire and I had to land. That was that for the race. We shipped the plane back to Los Angeles and had the engine overhauled."

The plane looked much like Lindbergh's in that the pilot also flew from the rear. The huge gas tank in the forward fuselage restricted vision just as it had done in the *Spirit of St. Louis.*

Meanwhile work proceeded on the Ryan Mechanics' completely new design. McNeel's idea was to build a 37½-foot span full-cantilever monoplane — not a strut braced job like the Ryan M-1, the *Spirit* and the 'rebuilt Cruzair'. It would be of all-metal structure, using welded chrome molybdenum seamless steel tubing even for the wing spars and ribs instead of the conventional wooden spars and ribs of that period. Power plant would be a 200-horsepower Wright J4-B Whirlwind radial engine.

While the design and metal construction concept of the CM-1 (cantilever monoplane No. 1) was McNeel's, the preliminary engineering was done by Otto Timm and the final detailed engineering and stress analysis by Waterhouse.

With much Hollywood fanfare, the 5-place *Lone Eagle,* with McNeel at the controls, made its first flight from the Western Air Express terminal — Vail Field — on Telegraph Road in Los Angeles on February 12, 1928.

From Bogota, Colombia, Lindbergh had wired his good wishes to "Mac, Henry and Fred," together with permission (though none was required) to use the name *Lone Eagle.*

Lone Eagle builders headed by O. R. McNeel, president, center. Henry Hunold, vice president, on McNeel's left; Fred Ayars, secretary-treasurer, in business suit

Ryan Aeronautical Library

More than good wishes were needed, however. The airplane proved to be very tail heavy. Subsequently, 12 inches were added to the fuselage forward of the wing to compensate for the tail-heavy condition. And the power plant was changed to a Menasco Salmson 260 h.p. engine.

Because of confusion caused by the use of Ryan's name by the Ryan Mechanics Monoplane Co., McNeel and his associates changed the name of their budding firm. It became Federal Aircraft Corporation ("an expansion of Ryan Mechanics Monoplane Co.") which they incorporated February 14, 1928 in Los Angeles with an authorized capital of $200,000.

Unfortunately sufficient funds were not available through their financial backer, B. A. Seitz, to continue the Federal operation in Los Angeles. However, Seitz had business friends in nearby San Bernardino who with Chamber of Commerce backing were interested in getting into airplane manufacturing.

There, Federal Aircraft built another version, the six-place Wasp-powered CM-2 in which the design changed from the all-metal cantilever structure of the *Lone Eagle* to a strut-braced model of the same high-wing monoplane. In early flights of the CM-1 *Lone Eagle,* McNeel had difficulty with the hinged canopy and this feature was replaced in the CM-2 by a 'conning tower' type windscreen.

Following the CM-2 came the Federal three-place-in-tandem CM-3 cabin monoplane built in a packing house at San Bernardino, again with engineering assistance from William J. Waterhouse. In appearance the CM-3, designed around the seven-cylinder Hallett 130 h.p. engine, was very similar to the Curtiss Robin. Later it was powered by an eight-cylinder Kinner engine and bore license X-7667. Like the *Lone Eagle,* the entire structure was of welded chrome molybdenum steel tubing. Lon E. Wheeler from the old Ryan Airlines crew at San Diego also joined Federal at this time.

In a "Hello, Charlie," letter to Lindbergh, Fred Ayars wrote that "We took your advice and got a new name and are building the prettiest little three-place enclosed job you ever saw.

"We've just finished raising our first $50,000, which is always the hardest, and are going after about another hundred thousand preparatory to building a new factory and buying a field and getting started right."

Money to meet the accumulation of bills was getting hard to come by, for Ayars had to write Waterhouse on October 11, 1928 that "I have been stalling writing you for some time hoping to get your other check for $300 but it looks like another week or ten days before I can make it." With no product going out the door, the problem of getting funds to meet expenses was getting critical.

One of McNeel's friends from the early days at Rockwell Field, San Diego, where both were civilian pilot-mechanics, was C. R. (Dick) Bowman. Later both worked for Claude Ryan. Bowman flew the Los Angeles - San Diego Air Line daily passenger route for Ryan Airlines, Inc., then went to Pacific Air Transport as their pioneer pilot introducing the Ryan M-1 monoplanes on routes along the Pacific Coast.

Taking time off from his PAT piloting duties, Bowman did some flying in the *Lone Eagle* and CM-3, then joined up with Chicago business man Chuck McGrew, reportedly film star Jean Harlow's first husband, to become special sales agents for Federal under Fred Ayars as the company's General Manager.

Bowman-McGrew had a contract for national distribution of the CM-3. They claimed to have a potential 50-plane order for the new monoplanes from an oil tool executive but he wanted a 51% ownership in Federal Aircraft. The deal was never consummated.

With aircraft manufacturing companies popping up all over the country, there were just more new airplanes being offered than the public was able to absorb. Federal Aircraft came on difficult days. Its goal of turning out a $5,000 plane a day by January 1, 1929 was illusionary.

The stockholders felt it was unrealistic to continue, and they no doubt were incapable of financing the work-in-process. Things came to a grinding halt as operations were suspended in mid-1929. Fortunately McNeel recalled, "There was sufficient money and equipment to pay off stockholders 95 cents on the dollar of their original investment."

Reportedly, a delay in obtaining a Department of Commerce engineering approval of the plane's design blocked continued production and after some months Ayars left for work in Mexico.

The Waterhouse Cruzair which McNeel had used in the aborted trans-continental flight had gone to San Bernardino where it was used for short trips on company business. McNeel recalls that "When the company was dissolved, the plane was sold by Bowman-McGrew and the receipts used to help pay back the original stockholders. That's the last I saw of the Cruzair." No one knows its ultimate fate.

CM-3 cabin monoplane was final product of Federal Aircraft, successor of Ryan Mechanics company

Charles O. Ayars

WILLIAM J. WATERHOUSE

The Waterhouse Cruzair monoplane of 1926, whatever its aerodynamic merits, came upon sad days financially. Associated with William J. Waterhouse in Waterhouse Aircraft, Inc., of Glendale, California, was Lloyd Royer, formerly a welder with the Kinner company.

The first Cruzair, which cost about $2,000 to build, exclusive of the engine, was finished early in 1926. But, said Royer, they had no money for the Wright J-4 Whirlwind engine they wanted to install. Finally the Cruzair did fly, about two month's after Claude Ryan's M-1. They had, of course, hoped to sell the Cruzair to Vern Gorst's Pacific Air Transport but Claude Ryan beat them to it. Royer stated that the Cruzair proved to be an excellent airplane but that there were no commercial sales, at least none of consequence.

Waterhouse, however, did sign up Klamath Air Service Company of Klamath Falls, Oregon and Corning, California, as distributor and there was an apparent attempt to consolidate the Waterhouse and Klamath businesses.

With no money coming in, Klamath had to sell its Wright-powered Cruzair and a Roamair (minus engine), another Waterhouse-designed plane, to satisfy a bank loan.

Waterhouse and H. H. Tharp, who had been the organizer of Klamath, later went to Oakland, California, and in the back shops of the Robison Tractor Company they began work on a new radial air-cooled aircraft engine Waterhouse had designed.

In October, Waterhouse returned to Southern California and opened up in business as a consulting Aeronautical Engineer operating out of an office in Hollywood. Rather than start a new manufacturing company of his own, Waterhouse decided to concentrate on providing engineering service to entrepreneurs trying to bring their own aircraft and engine concepts to fruition.

Waterhouse was certainly not lacking credentials, nor the ability to get across to prospecttive clients his capabilities. They were told that "Mr. Waterhouse built his first plane in 1910" and that it was "very similar to the Wright machine.

"There being no fully accredited flying schools at the time, he had to teach himself to fly. The next plane was a monoplane with the pilot seated below the wing on a strip of canvas and controlling the ship with an inverted stick control made from an old umbrella handle.

Robert Peck

Bill Waterhouse built his own Continental-powered monoplane in 1910 and taught himself to fly

"The lack of engines, sufficiently light, was the most serious handicap at that time. To overcome this difficulty, Mr. Waterhouse with the help of a young friend, designed and built their own engine. It was also necessary to design a carburetor because there were none on the market for an engine of this size. For several years it was on the market under the name of Waterhouse carburetor.

"Later, Mr. Waterhouse was Assistant to Dr. A. F. Zahm at the Curtiss Aeroplane and Engine Corp. of Hammondsport and Buffalo. During the War he was Chief Inspector at the Standard Aeroplane Co. and Aeronautical Engineer with the U. S. Air Service inspection department at McCook Field.

"Subsequently, Waterhouse went to California where he became an associate engineer with the Douglas-Davis Co., while the first Douglas airplane, the *Cloudster* was being designed and built."

One of the Waterhouse's new consulting jobs was for Tijuana Aircraft Co., of which Gov. Abelardo Rodriquez of Baja California was the head. The Mexican government was anxious to design and build its own aircraft and so helped set up the new company just across the border from San Diego.

Two monoplanes were completed early in 1928 from plans furnished by the Mexican War Department and adapted by Waterhouse. April 5 when Col. Charles A. Lindbergh was in San Diego to take delivery of his new Ryan B-1X, three of his associates who backed the trans-Atlantic flight, visited the Tijuana aircraft plant.

The planes were approximately the size and general appearance of the Ryan Brougham. In test flights of the first plane, an observation type, a six cylinder German B.M.W. engine of 185 h.p. was used, but the second plane, expected to carry 380 gallons of fuel and make a non-stop flight to Mexico City was scheduled to have a Wright Whirlwind.

As mentioned in a promotional bulletin from Waterhouse Consultants, "Mr. Waterhouse recently designed and supervised construction of a military model for the Mexican government which was named the *Baja California* and which made the highest speed, non-stop record from this country to Mexico City, averaging 108 miles per hour on the flight from Mexicali."

Even as Waterhouse was consulting on the Mexican aircraft, he was also busy in the role he had taken in January 1928 as dean of the Western College of Aeronautics on Slauson Avenue in Los Angeles. This arrangement was hardly a satisfactory one as the school could not meet its financial commitment to Waterhouse and they parted company under somewhat strained relations within the year.

Another client was the Courier Monoplane Co. of Long Beach, California whose Waterhouse-engineered three-place monoplane made its first flight February 12, 1929. The Courier cabin plane looked much like most strut-braced monoplanes of the post-Lindbergh era with a strong resemblance to the Federal CM-3 and Curtiss Robin.

Early in 1929 Waterhouse entered some type of agreement with Earle Ovington probably in connection with the "Roamair" plane. Ovington, of Santa Barbara, California, had made a name for himself in aviation in 1911 by carrying mail by air on Long Island.

In a later letter to Ovington, Waterhouse released his trademark claim to Ovington for use of the name "Roamair" for "your airplanes." "I originated this name," Waterhouse wrote, "and first used it for a biplane I designed and built at Glendale about four years ago (1925)"

Inquiries for engineering help continued to flow into the offices of William J. Waterhouse and Associates. A typical one in April 1929 was from a San Francisco firm "contemplating entering the airplane manufacturing business" which would like to "Have our drawings checked and stress analyzed for approved type certificate."

Two weeks later Diablo Aircraft Company of Stockton, California advised Waterhouse they were planning to incorporate to build an airplane.

"We intend," Diablo wrote, "to design the two-place monoplane as far as main dimensions, general principles, features and other parts (are concerned). What we wish you to do is to check and finish design, get stress analysis and approved type certificate."

A month later Waterhouse was writing Dr. Alexander Klemin of New York University to advise that "I am designing for a Mr. Cornelius a high wing monoplane type airplane." Concerning the design Waterhouse said, "I am of the opinion that Mr. Cornelius had incorporated many ideas worthy of development, particularly the principle of coordinating the function of the wings with that of the tail surface group; also the clever manner in which he had worked out the operation of the control mechanism."

Virtually every Waterhouse consulting arrangement ran into difficulties.

Although Waterhouse was no longer involved with the former Tijuana Aircraft Co., his *Baja California* design was plagued with troubles.

The plane that the company had constructed for the personal use of the Minister of War, General Joaquin Amaro, had a forced landing somewhere in the State of Jalisco and suffered serious damage.

Senor Jose Flavio Rivera obtained information that the plane, *Baja California No. 3,* was lying abandoned in the military shops at Valbuena Field in Mexico City.

Hawley Bowlus

Waterhouse Cruzair, a near duplicate of Ryan M-1, never was able to compete successfully with its rival monoplane

Rivera organized Compania Aeronautica Nacional, S.A., in Tijuana, and obtained a special permit from Emilo Portes Gil, President of Mexico, to repair the aircraft. It was at this time, after the break-up of Federal Aircraft, that Fred Ayars, one of the organizers of Ryan Mechanics Monoplane Co., joined Rivera's project.

(*B.C. No. 3* which had the familiar Waterhouse tail group — appears to have been a rebuilt or follow-on to the planes built earlier by Gov. Rodriquez' Tijuana Airplane Co. Col. Robert Fierro is reported to have been the pilot of *B.C. 2* which, according to a San Bernardino newspaper in August 1928, "is being built by the Tijuana, Mexico, aeroplane company owned and operated by former associates of the Federal Aircraft corporation heads."—Ed).

"*B.C. 3* was shipped to Tijuana by railroad and Rivera was granted three months to repair it," according to El Piloto Magazine, "then it was to make a flight across the Republic in behalf of the national aviation and the anti-alcoholic campaign.

"Things did not turn out as first planned and an extension of time was solicited, but at the end of the fourth month the Federal Government allowed only a certain amount of additional time for the aeroplane to be returned to the hangars at Valbuena and there be sold at public auction.

"Sr. Rivera obtained via telegraph a last grace of only a few days and a permit to attempt a non-stop flight from Tijuana to the Republic of San Salvador, in Central America.

"During the long period of time used in the repairing and conditioning of the craft, various Mexican workers helped Sr. Rivera who counted among his helpers a young North American expert by the name of Fred C. Ayars. This young fellow, kind, intelligent and active, worked unceasingly day and night, making a careful inspection and rectification of all the parts of the aeroplane.

"The work on the aeroplane was nearly completed when another grave difficulty appeared; a pilot was needed. In the end, Fred managed to get a friend of his, pilot-aviator Benny Moeur to accept the piloting of the aeroplane on the non-stop flight. (Young Moeur had been a test pilot for Federal aircraft. Four years later his father, Dr. B. B. Moeur, became Governor of Arizona.)

"At last the day arrived, the first of November, 1929, that had been decided for the flight.

"There on the Playa (beach) of Rosarito was the *B.C.3* with its enormous white wings and a black arrow, like a fatal sign, that 'Pop' Ayars himself had painted along the fuselage.

"The aeroplane would be piloted by Moeur and, as mechanics, there would go along Ayars and Alberto Figueroa, a young Mexican.

"A tank in the middle of the fuselage served as a seat for Mechanic Figueroa. Ayars had to lie on top of the main gas tank in a horizontal position where it was impossible for him to make any quick movement.

"The powerful 'Wasp' made its four hundred horses neigh. The propeller tuned its deathly chant. Inside the heart

Charles O. Ayars

"Baja California 3" in which Fred Ayars of Ryan Mechanics Monoplane Co. was killed on attempted long distance flight

of the craft were 570 gallons of gasoline. The craft responded with the roar of an apocolyptic beast and started dragging itself like an infuriated dragon. The heavy craft was unable to take off.

"One time and another, and yet another they tried to initiate the flight. Ninety-five miles an hour . . . a jump of the craft . . . then another. Now it is leaving the ground. But all of a sudden a tire comes off one of the wheels. In front there is a ditch. The aeroplane touches the ground again for the last time.

"Violently it strikes on one of its wings, the propeller and engine hit the ground breaking into a hundred pieces. Inverted, the apocolyptic bird, like a whipped colossal, strikes its back against the ground.

"Figueroa did not suffer the least injury. The pilot, after 15 days in the hospital returned to his work. Taken from the wreck, Ayars was rushed in an automobile (which ran out of gas on the way—Ed.) to the nearby American town of National City. Under the stress of delerium 'Pop' wanted at any cost to be taken anew to Rosarito to try the flight again. But he outlived the catastrophe by only a few hours and died in the hospital."

For the Courier Monoplane Company, Waterhouse was to have done the engineering for both the three-place PR-1 cabin monoplane and the two-place open-cockpit MT-1 trainer. By January 1930, Courier was $1,200 behind in meeting the $3,500 contract price to be paid Waterhouse and he had written them that "our only other chance of obtaining payment for that design is to reclaim complete ownership of the design in order to be free to sell it to some other company."

The economic reins were tightening on Waterhouse. Even Western Flying magazine was unable to collect from Waterhouse for advertising in that trade journal, and the account had been turned over to a collection agency.

The great Depression of the thirties brought more problems. By 1935, Waterhouse's back was to the wall as he wrote a collection service. "I am still unemployed and am dependent on my friends and relatives for food and shelter. There have been times when conditions were so critical that application for county relief has appeared evident. However, in spite of my adverse experiences during the past three years, I still have faith in the future and my ability to meet my obligations."

The last file which turns upon on Waterhouse is a July 9, 1940 letter from Manta Aircraft Corporation of Los Angeles concerning his pre-subscription agreement to buy 100 shares of stock at $10 per share. One report has it that Manta was formed by David R. Davis (who had been partner with Donald Douglas in the original Davis-Douglas Company) and Waterhouse to build the Davis design of a Manta-shaped airplane. Because of World War II, the project never got off the ground.

Clearly Waterhouse's talent lay more in the field of engineering and stress analysis than as an innovator of design concepts or business management.

"Good ideas," he had written, "are being developed which in most cases only need the practical technical advice of an engineer to make the activity a success." If bringing a lot of other people's ideas to fruition was a suitable yardstick then Waterhouse was indeed a success as an aeronautical engineer.

DONALD A. HALL

Donald Hall, the near-recluse engineer whose high standards of workmanship and quality had so much to do with the success of Lindbergh's trans-Atlantic venture, never again — unfortunately — hit the pinnacle of professional attainment he achieved in his design work on the *Spirit of St. Louis*.

His X-1 *Doodle Bug* — with flying tail and gear shift — was an innovative approach which, however, did not meet the needs of the aircraft market of that day. Yet Hall considered the design concept his most important contribution to aviation.

Like many others who had important roles in building Lindbergh's plane, nothing seemed to go right after that for Don Hall. It wasn't that their careers were jinxed only that the stimulus of such an event could not be topped or even repeated.

Always withdrawn and a loner, Hall became even more so with the passing of the years. His death from a heart attack on May 2, 1968 was a secret well-kept from the public by his family. Funeral services were private and no word of his death leaked out until nearly two months later when his son visited the writer.

We had written Don to ask clarification about the relationship between the first Ryan Brougham and the *Spirit*. Hall had written out answers to our questions and given them to his son Don Hall, Jr. In late June the son delivered the answers and a covering letter he had written. It read, "I am writing this letter instead of my father who died suddenly from a heart attack in early May."

It was the first anyone outside the family knew of Don's passing.

Several days later Lee Dye, a reporter for the San Diego Evening Tribune, was to write:

> The fleeting finger of fame touched his life briefly and then moved on, and when he died in his home here a world which had sung his praises scarcely knew he was gone.

Things looked bleak for Don Hall in September 1928. His X-1, which he said would "test several more or less radical principles," was his great hope but the new St. Louis owners were "devoting all their energy to the production of the Brougham, already obsolescent."

Hall also sought Lindbergh's help. "There is much I would like to discuss with you in this connection," he wrote when sending Slim a copy of the letter to Guggenheim, "and would greatly appreate it if you could plan to spend at least two or three days here."

Then Hall struck out on his own, planning to complete the development of the X-1 on which he held the patent rights. But aircraft research and development required money, and Hall had little. For about two years, he work and lived at the YMCA, then found he lacked funds to maintain even that level of existence.

With his small reserve of cash depleted and no backer of consequence to underwrite his project, Hall arranged to live and work in his old Ryan Airlines office in the now deserted fish cannery.

Wilhelm Schult

Donald A. Hall devoted his post-Lindbergh attention to development of radical new monoplane

H. A. Erickson

Mahoney-Ryan X-1 Special, the Hall designed monoplane which, unfortunately, did not meet the needs of the day

He placed a cot and hot plate on one side of the drafting room. He really had to rough it. He washed his clothes in a bucket and hung them in the wing loft to dry. He cooked his meager meals on as little as 10 cents a day by buying only the cheapest foods.

Hall's fame was almost forgotten during the early thirties. But on May 13, 1931 Hall and sixteen friends invested a token $10 each to purchase the first 17 shares of the new Donald Hall Aeronautical Co., Ltd., ('Custom Built Airplanes') capitalized at $200,000.

One of the partners was a bank executive to whom Lindbergh had written a non-commital letter a few weeks earlier reassuring him that "I have great confidence in Hall's ability as an aircraft engineer and know of no one whose opinion I value more highly in this field."

Months earlier the banker had written Lindbergh soliciting his endorsement "As you know" he wrote, "Don is an idealist, and lacks the experience necessary to talk finances with possible investors. He has waited for someone to take charge of the matter, and his personal means are nearly exhausted. The bank has talked to a few local men who have been only passively complimentary toward the new ship, and as you know, half hearted praise is no recommendation."

Handicapped though he was in getting his ideas accepted Don Hall did come up with several new aircraft designs which he offered prospective customers under his new corporate banner.

One design was the XS-1B Super Speed Mail-Express "using the Hall controllable aft (elevator) wing and a new principle of wing construction permitting the chassis wheels to be partially enclosed within the wing." Its calculated maximum speed was to be 242 miles an hour, twice that of the *Spirit of St. Louis.*

The prospects for using his controllable aft wing for the newly designed plane were no brighter than for the original X-1 *Doodle Bug,* yet Hall continued to try and interest Lindbergh and his influential friends. But in a letter to Col. Henry Breckenridge, Lindbergh's attorney, Hall admitted that "I have been plugging along with my work without any more success than considerably increasing our patent and design assets with practically no expenditure."

This he explained in a letter to Lindbergh's good friend, Harry F. Guggenheim, sponsor a year earlier of Slim's nation-wide tour in the *Spirit of St. Louis* to promote interest in aviation.

Hall envisioned a "purely experimental corporation here in San Diego," to continue work on his concepts. "I am confident," he wrote Guggenheim, "that this proposed airplane development corporation can be successfully organized and run for a year with rigid economy for $100,000.

"There is need for immediate action since the Mahoney Company is to practically close down the San Diego plant in two weeks and I would like to recruit two of my most efficient associates for this new corporation."

Hall did find occasional work as a consulting engineering for others. Claude Ryan had remained in San Diego after the Lindbergh flight and in 1932 was doing business under the name Ryan Aeronautical Company. Always full of new design ideas, Ryan had Hall in late 1932 draw up preliminary three-view drawings for a tailless Two Seater Sport Coupe and for a military Expeditionary Pursuit, but these concepts never got beyond the drafting board.

The City of San Diego took over the old cannery building but allowed Hall to remain in his old drafting room until Solar Aircraft moved in, in1933. Most of the time he lived there rent free. He couldn't sell his only possession of any value, a 1928 Chrysler roadster because it was collateral for an unpaid loan. Occasionally he would splurge and buy one gallon of gasoline to take his girl for a ride, but usually he walked or rode a streetcar.

A new business partner was finally found in 1933. The partnership gave him some measure of security, so he married the San Diego girl, Elizabeth Walker, who had waited for him since 1930. But reverses plagued the business partnership, and early in 1936, it was dissolved and he was once more without work.

The year 1936 found Hall living in a small furnished house and wondering what the future held for him, his wife, and their one-year-old son, Donald, Jr. Looking back upon the meager years, Hall resolved to put his dream aside and go to work for an established company.

He started out in 1937 with Consolidated Aircraft Corporation, serving first in his profession as an engineer, then in a new specialization, patent analysis. He was the first Patent Director for Convair.

When a slump hit the aircraft industry in 1949, Hall took a leave of absence and worked a short time for a Los Angeles firm. Then in December 1949, he was hired as an aeronautical engineer at the Naval Air Station, North Island. In June 1952, he was selected to head the helicopter engineering unit of the overhaul and repair division, remaining in that post eleven more years until poor health and his 64 years forced his retirement in 1963.

A man of integrity and high standards of professional engineering, Hall — despite his lack of business success — was always held in especially high regard by Lindbergh.

When hearing, belatedly, of Hall's death, Lindbergh wrote his widow that "The *Spirit of St. Louis* hanging there in the National Museum (Smithsonian Institution) will always be a memorial to his fortitude and genius."

Ryan Aeronautical Library

Old friends Lindbergh, Hall and Dan Burnett meet 15 years after they worked together on "Spirit of St. Louis"

B. FRANKLIN MAHONEY

Perhaps the most succinct, accurate description of Benjamin Franklin Mahoney was given forty years after the dramatic events which followed Lindbergh's famous flight. In a letter to the writer, Tom Mathews, the company's former publicist, wrote —

"Mahoney may have lived the Roaring Twenties somewhat in the F. Scott Fitzgerald manner but he was a pretty smart Irishman who, unfortunately, had more than his share of human weakness."

Like many others, the high point of Mahoney's life was the success of the *Spirit of St. Louis* and the recognition all associated with the New York-Paris flight shared in some measure with Lindbergh, the incomparable pilot. Thereafter they were unable to cope with changing conditions, retreating instead to past glories and reliving those stimulating days.

But Mahoney, for one, already had a head start on personal problems.

He had come out to San Diego in 1919 with his widowed mother, Mrs. Jennie Mahoney, whose husband had owned a chain of prosperous retail stores in Pennsylvania. The next year Franklin married Helen Ann Post and went to work in San Diego as a bond salesman but gave up that employment in mid-1923. Meanwhile his mother purchased several parcels of downtown real estate.

Soon after the birth of their second son, Franklin and his wife came to a parting of the ways. An article in the San Diego Sun of July 24, 1924 hinted at some of the matters detailed in the divorce complaint:

TO FIGHT WIFE
Mahoney to File Counter
Charges to Divorce
Allegations

"A cross complaint for divorce and a general denial of charges contained in his wife's divorce action, will be instituted by B. Franklin Mahoney, who recently accused Edward H. Post, his father-in-law, with attempted murder, it was announced by Mahoney's attorney today. Under the law, 10 days is allowed for the filing of the return action.

"Preliminary hearing on the divorce complaint with the view of allowing Mrs. Mahoney temporary maintenance during the progress of the final action, if she would be destitute without it, will be held before Judge Marsh next Wednesday.

"We will contest every allegation and claim made by Mrs. Mahoney," William Mirow, attorney for the defendant, said today.

"Alleging cruelty, drunkenness and undue familiraity with other women, Helen Ann Mahoney filed the first divorce suit."

A month later the case was dismissed but the couple agreed to continue to disagree. For diversion, Franklin went out to Ryan Field on Dutch Flats and learned to fly under Claude Ryan's tutelage. His mother, who could provide financial backing, thought his interest in aviation might settle him down.

Mahoney was full of ideas and enthusiastic about the future of aviation. He played the role of risk-taker in contrast to Ryan's conservative approach to business. In April 1925 they formed a partnership to start the Los Angeles-San Diego Air Line and the Mahoneys moved to Los Angeles for a period so Franklin could handle the airline business at that end. But in time the relationship with Ryan also went by the board and Claude sold out his remaining half to Mahoney in November 1926.

Six months later came the success of Lindbergh's flight, the formation by Franklin Mahoney of his own firm and a second marriage, this time to "Tommy" with whom he had been keeping company.

With the plant moving to St. Louis late in 1928 under its new ownership, Franklin and Tommy also went east. Mahoney, and presumably his wife, were in Europe when newspapers carried the word that he had sold all his holdings in the company and had given up his management role in the St. Louis-based business.

The Mahoneys settled down in an apartment on Park Avenue in New York where they reportedly entertained on a lavish scale. Ever the gambler, Mahoney invested heavily in the stock market. He'd always been lucky in the business risks and gambling chances he'd taken but this time the market crash of October 1929 swept over him as it did so many.

Attorney William Mirow, center, hosts friends B. Franklin Mahoney and Charles A. Lindbergh

William Mirow estate

From time to time Mahoney popped up here and there in aviation circles, but the trail was hard to follow. In 1931 he was living in the loft of a hangar at Roosevelt Field, "down and out" according to a pilot associate from San Diego who was shocked to see his condition. "I don't know," he related "when I have ever been quite so upset by something."

A year later, friends were writing Lindbergh because of their concern for Mahoney:

> "Tommy" Mahoney was at our house for dinner last night and she told me . . . of her getting an order from you for stationery. You have no idea how your action pepped her up, and she is full of additional enthusiasm for her job.
>
> I am doing what I can in my quiet way to get her contacts with big companies so that she may build up a substantial business.
>
> I learned from her that B.F. is pretty well 'down and out', and if this is the case I am rather sorry that he hasn't been in to see (us).
>
> I am writing you because "B.F." may be in to see you again and if he does drop in, why don't you suggest that he come up to see us? If his condition is as sad as I have been led to believe, I think it is up to us who really think we are his good friends, to endeavor to straighten him out and get him back on his feet. If he needs a little money, I shall be glad to do what I can personally.

Some time later, Mahoney went out to Wright Aeronautical at Paterson, New Jersey, to look for employment. His former public relations man, Bob Adams, was by then head of advertising for Wright, but was unable to give Mahoney any help as Franklin had little knowledge of aircraft engines.

In December 1937, Mahoney's name was again in the news. After ten years of marriage, "Tommy" filed for divorce in New York City closing another phase in Franklin's kaleidoscopic career.

The next year, on the eleventh anniversary of Lindbergh's flight, a newspaper reported that Mahoney was serving as sales manager for a company Hawley Bowlus, former Ryan factory superintendent, had formed in the east to build sailplanes and auto trailers.

Some years later Mahoney remarried in Los Angeles and went to work as a salesman for Vultee Aircraft. Later he sold airport lighting equipment but in time his health began to fail. On July 31, 1951, at the age of 50, Benjamin Franklin Mahoney passed away in a sanitarium near Los Angeles having spent nearly eight years there with a heart ailment.

Sensor Studio

Benjamin Franklin Mahoney . . . "my good Irish friend"

Once, when asked to confirm reports that Mahoney often gambled in Tijuana in order to meet the company payroll, Lindbergh replied "it would have been in keeping with the character of my good Irish friend, Benjamin Franklin Mahoney." It was a fitting tribute to the likeable nature of a very generous and outgoing man who, unfortunately, lacked the business ability and rigid self-discipline to capitalize on the unique opportunity of having owned the company that built Lindbergh's plane.

A. J. EDWARDS

Sensor Studio

A. J. Edwards . . . boy, how he lived!

Like many others prominently identified with the Lindbergh flight, A. J. Edwards, the Ryan Airlines sales manager, came on sad days after that zenith of his career when he was Slim's self-appointed 'manager.' A.J. was particularly conscious of his role because Lindbergh, after a short stay at the U.S. Grant Hotel, shared Edwards' apartment in San Diego during construction of the *Spirit of St. Louis.*

Edwards had early achieved some notoriety as a race car driver who challenged such pilots as Lincoln Beachey in airplane vs. auto races at county fairs. He had long been an automobile dealer in the northwest and later in the midwest and was the principal aviation enthusiast of Muskogee, Oklahoma, before moving to San Diego in 1926.

At Ryan Airlines, Inc., and the successor B. F. Mahoney Aircraft Corp., the flamboyant Edwards lasted only a year. But during that time he cut a wide swath. Business responsibilities for which he was ill equipped were forced upon him but, as his son Clarence R. (Curly) Edwards recalled, A.J. "knew how to live high."

"One time he made $15,000 cash on a deal, and it lasted him only three weeks. It wasn't anything unusual. He could live on $5 or on any amount large or small. He felt money was to be made and spent, not accumulated. He could never see any advantage in tomorrow. He lived today, and boy how he lived!"

He considered himself, and apparently was, quite a man with the ladies. Even son Curly admitted "I never had an accurate check on how many times he was married."

With the sale of Mahoney's company to St. Louis business interests, Edwards decided to remain in San Diego where in January 1928 he joined the newly formed Prudden-San Diego Airplane Co. as sales manager. George Prudden, formerly an engineer with the Stout Division of Ford Motor Co., had helped engineer early Ford metal planes and was now introducing a similar design of his own.

Things, however, did not go too well in the new company. By June, Edwards was wiring Lindbergh for help.

PLANE SIX HUNDRED POUNDS OVERWIGHT, CUTTING DOWN PERFORMANCE TO SUCH EXTENT AS TO TEMPORARILY CANCEL ORDERS AND CLOSE PLANT SUBJECT TO REORGANIZATION. I MAY BE OPEN TO CONSIDER PROPOSITION IF YOU KNOW OF ANY.

Ten days later, again to Lindbergh:

HAVE JUST TERMINATED MY CONNECTION WITH PRUDDEN COMPANY ACCOUNT FACTORY CLOSING. WOULD LIKE TO MEET YOU AND HAVE DISCUSSION BEFORE I MAKE ANOTHER BUSINESS CONNECTION IF YOU WILL ADVISE ME TIME AND PLACE AGREEABLE TO YOU.

And six months later from Wichita, Kansas, he wired again advising that he was the new general manager of Swift Aircraft and inviting Lindbergh to "drop in when you come through."

In the Fall of the following year, 1930, A.J. wrote "Dear friend Charles" reminding him "that if the time ever came when you could do anything for me, you would be glad to do so.

"Since that time circumstances have so changed the prospective that I had hoped it might not be necessary for me to trouble you. Of course as you perhaps know, I seem to have been able, with the exception of the Ryan Company, to pick a lot of losers.

"In my desire to get off the road and get settled down I took what cash I had saved up this year, and invested it in a plane and the necessary expenses to become the Aeronca dealer here in Los Angeles.

"After fighting with them (an aviation school) for two months they have flatly refused to pay for the plane. This makes it necessary for me to seek a new connection.

"No doubt with your many connections you will know of a place where my services and experience would be of value to someone and in turn would give me a much needed connection. I wonder if I am asking too much of you to suggest your giving this a thought and letting me hear from you."

Much of Edwards' correspondence was to or through Lindbergh's attorney, Colonel Henry Breckenridge in New York City. In a plea to Breckenridge to get Lindbergh's help, Edwards stooped to sarcasm: "I presume he is too busy to even remember me now . . . and am just beginning to realize that I had very little to do with it (the flight) except all the work and the responsibility of getting the ship through on time."

Like many others, Edwards had deluded himself into thinking that Lindbergh would never have made it to Paris without some unique assistance which only he could give.

The Fall of 1942 found Edwards back in the automobile business in San Diego as a salesman. The Depression was not treating him kindly as described in yet another "Dear Friend Charley letter:

"I feel that if you knew how bad I need work you would be glad to suggest something I can do.

"I haven't heard from Mahoney for some time but I guess he is among the rest of us as far as money goes.

"When I came down here I moved into the old apartment where you and I lived before you went across. As no one is buying cars now I am having to move out this week. There are lots of old memories there though. Of 24 salesmen only two of us have delivered cars in 4 weeks, and I have one of them. Made $20. Can you beat that? I am sending this care of Col. Breckenridge hoping that you will get it."

The following year came a final plea to Colonel Breckenridge: "It always has seemed to me Charlie could sometime or other remember I have to live and through his excellent contacts obtain something for me to do of which I am capable."

Unfortunately for these concerned, Lindbergh could hardly be expected to be an employment agency for a group of fellow-workers, devoted as they were to him, whose relationship spanned a mere sixty days.

Some additional insight into the dashed hopes and aspirations of those connected with the New York-Paris flight had come from Howard Morin, a San Diego newspaperman who was among the first in the press to report about Lindbergh's trans-Atlantic plans.

He, too, had some disappointments — even while recognizing the human interest story possibilities. This was reflected in a January 22, 1929 letter he wrote to Edwards when A.J. was with Swift Aircraft in Wichita:

"At the suggestion of Harry Carr of the Los Angeles Times, I am going to submit to Liberty Magazine a story, the text of which substantially will be as follows:

"That while Lindbergh attained worldwide honors in his Paris flight, virtually every person even remotely connected with his achievement fell heir to misfortune or plain damn hard luck. In other words, you, Mahoney, Hall, Bowlus, the shop foreman, and even myself, lost our jobs.

"I intend writing the story, of course, without casting discredit on any of us. Carr thinks it will make a very dramatic story and so do I. Would you care if I mentioned that after the collapse of the Mahoney Company here you appealed to Lindbergh for assistance but as far as I am aware, you never received it with the exception of the promise made in his letter and which obviously never was fulfilled?

"If there are any personal items which you think should be incorporated in the Liberty Magazine story, please let me know.

"Everything here is hotsy totsy except aviation. That is as fleeting as a summer day at the South Pole."

When Edwards passed away in 1953 his death, like that of Donald Hall 15 years later, was virtually unnoticed.

Ready to leave San Diego for Paris, May 10, 1927. From left, Donald A. Hall, A. J. Edwards, Lindbergh and Lt. Col. Harry Graham

A. J. Edwards / H. A. Erickson

WILLIAM HAWLEY BOWLUS

Mrs. W. H. Bowlus

Master craftsman William Hawley Bowlus, first Ryan employee and many of products with which he was associated

A few days before Lindbergh's take-off from San Diego for St. Louis and New York, where he made final preparations for the flight to Paris, he posed for photographs with key Ryan Airlines people.

The select few included Mahoney, Hall, Edwards and William Hawley Bowlus. Among that group — and all the other workmen — Hawley was the true master craftsman.

Claude Ryan and Bowlus had been friends since 1918 when they learned to fly together. Five years later, after starting a flying service in San Diego the previous Fall, Ryan hired his first employee — Hawley.

It was Bowlus who supervised conversion of World War I Standard biplanes into cabin planes for Ryan's Los Angeles-San Diego Air Lines. It was Bowlus who completed building the M-1 after Bill Waterhouse failed to furnish all the engineering drawings needed for its construction.

It was Hawley who was factory superintendent during the building of the *Spirit of St. Louis.* With Hall and Lindbergh he shared most of the responsibility for the plane's success.

When Mahoney sold out to the St. Louis crowd, Bowlus went back to work for Claude Ryan who meanwhile had started a new company — Ryan Aeronautical Corporation — to import Siemens-Halske aircraft engines from Germany and market them in the United States under the Ryan-Siemens name.

Bowlus had long been interested in gliding and was soon building a sailplane of his own design in which he set an American soaring record in 1929 of 9 hours and five minutes. During a San Diego visit that same year by Lindbergh and his bride, Anne Morrow Lindbergh, Bowlus taught both to fly sailplanes at San Diego and at Lebec, in the Tehachapi Mountains north of Los Angeles. Anne Lindbergh, in fact, received FAI "Ladies License" No. 1 for glider pilots.

Interested in all forms of transportation, Bowlus in 1933 designed and built what was probably the first all-metal vacation trailer to use aircraft production techniques. It was innovative, way ahead of its day and on the highway forty years later would hold its own for streamlined design. To build the "Road Chief" he formed his own company in San Fernando, California.

Between 1936 and 1942, the Bowlus "Baby Albatross" sailplane was supplied, in kit form, to soaring enthusiasts all over the world. Then, during World War II, Hawley conceived and supervised construction for Douglas Aircraft of the huge XCG-16 cargo glider designed to carry 46 fully-equipped soldiers when towed behind a powered aircraft.

Bowlus' many-faceted talents showed up in his post-World War II employment with Hughes Tool Company in building precision models and prototypes of new products. Later with his own firm, Bowlus Engineering, Inc., he did similar work for other companies and developed products for his own manufacture.

Long interested in working models of railroad engines and complete trains, Hawley turned his backyard hobby into another business venture in which he was active from 1946 until his death. Model railroaders agreed that no one could produce more intricate, more perfect parts and equipment than Hawley.

But Hawley wanted one last fling at flying. Even though he hadn't flown for years he had earlier logged nearly ten thousand hours of powered and unpowered flying time. Ten days before his death in 1967 he trucked an Albatross sailplane out to the desert and was towed aloft. He hadn't lost the touch.

Hawley was ever the craftsman and innovator, but was never able to convert his ideas into a really successful business venture. But to him that was relatively unimportant. What mattered was the unique skill and craftsmanship which were his hallmark.

The story of the man who missed the greatest opportunity in aviation — because he had sold his interest in the company just before Lindbergh's flight—has already been told in "Ryan, the Aviator."

His career, his business success, was in marked contrast to the stories of other participants. At the time Claude Ryan appeared to have 'missed the boat,' but in the long run he seems to have been the only one who really caught it.

As he wrote Lindbergh many years later in recalling "that period of high excitement:"

"I was in the ironical position of being on the outside while the little company I had started, built up and operated until shortly before the flight, was suddenly in the middle of world attention.

"I am sure now that trick of fate was a most fortunate circumstance for me because otherwise, at the age of 29, I very possibly would have been deluded into believing I was something which I definitely was not."

Hall, Mahoney, Edwards, Bowlus. All are gone. But at 76, Claude Ryan, a remarkable man, was still going strong as these words were written.

© KARSH — Ottawa

T. CLAUDE RYAN

COLORING NOTES

OVERALL SILVER DOPED,
DETAILING BLACK.
ALUMINUM NOSE AREA, BURNISHED
CABIN WINDOWS AND DOORS,
STAMPED ALUMINUM SHEET

TAIL SURFACES METAL FRAMEWORK, FABRIC COVERED

GENERAL SPECIFICATIONS

WEIGHT EMPTY	2250 lbs.
USEFULL LOAD	1750 lbs.
(PAYLOAD	935 lbs)
GROSS WEIGHT	4000 lbs.
TOP SPEED	138 mph
CRUISING SPEED	112 mph
CRUISING RANGE	750 miles
LANDING SPEED	55 mph
CLIMB, INITIAL, S.L.	1000 ft./min.
SERVICE CEILING	18,000 feet
ABSOLUTE CEILING	25,000 feet
WING AREA	224 sq.ft.
WING LOADING	14.5 lb./sq.ft.
POWER LOADING	13.3 lbs/HP

ATC No. 142

POWER - WRIGHT J-6-9 "WHIRLWIND"
RATED 300/330 HP

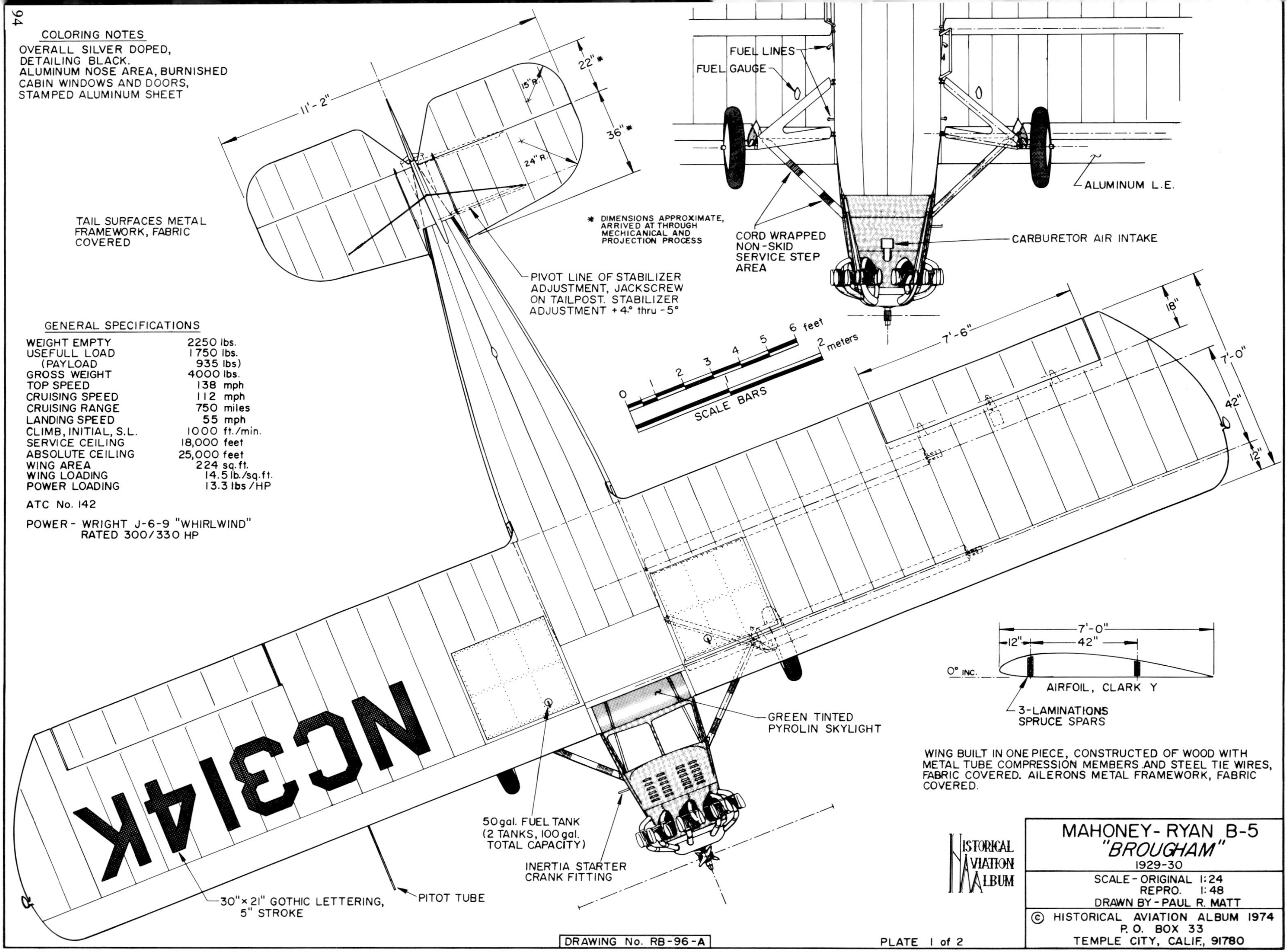

WING BUILT IN ONE PIECE, CONSTRUCTED OF WOOD WITH METAL TUBE COMPRESSION MEMBERS AND STEEL TIE WIRES, FABRIC COVERED. AILERONS METAL FRAMEWORK, FABRIC COVERED.

HISTORICAL AVIATION ALBUM

MAHONEY-RYAN B-5
"BROUGHAM"
1929-30

SCALE - ORIGINAL 1:24
REPRO. 1:48
DRAWN BY - PAUL R. MATT

DRAWING No. RB-96-A

PLATE 1 of 2

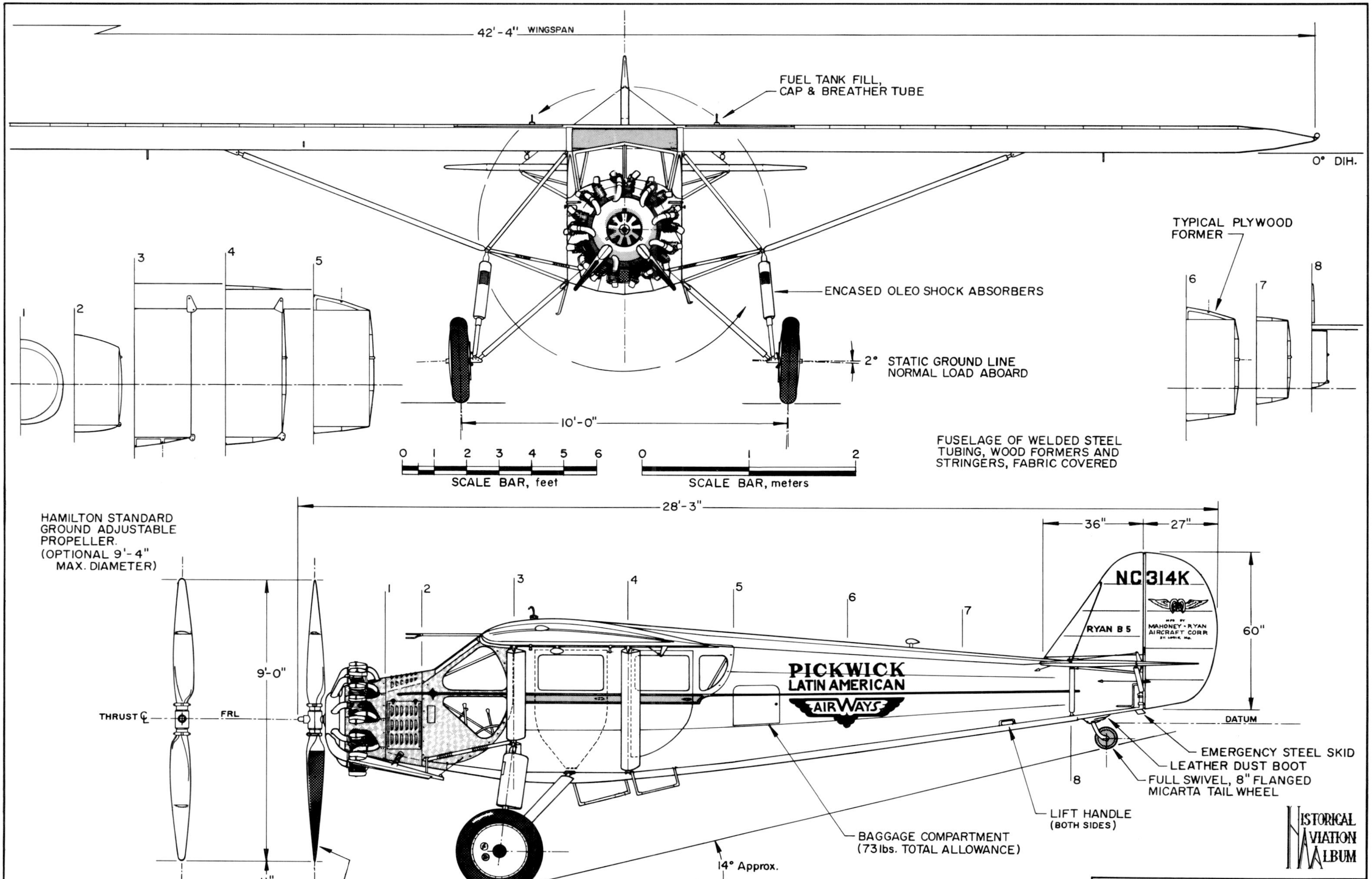

MY APPRECIATION IS EXTENDED TO WILLIAM WAGNER AND THE RYAN AERONAUTICAL LIBRARY FOR SUITABLE RESEARCH MATERIAL SO ESSENTIAL IN THE PREPARATION OF THESE DRAWINGS

Paul M.

MAHONEY-RYAN B-5 *"BROUGHAM"*

1929-30

SCALE - ORIGINAL 1:24
REPRO. 1:48
DRAWN BY - PAUL R. MATT

DRAWING No. RB-96-A

PLATE 2 of 2

PERFORMANCE AND SPECIFICATIONS — RYAN BROUGHAM AIRCRAFT

Model	B-1	B-3	B-5	B-7	C-1
A. T. C. No.	25	104	142	262	346
Builder	Mahoney	Mahoney-Ryan	Mahoney-Ryan	Ryan Aircraft (Detroit Aircraft)	Ryan Aircraft (Detroit Aircraft)
Serial Nos.	29-178	179-186	187-248	249-256	400-402
Number Built	150	8	62	8	3
Engine	Wright J-5-C	Wright J-5-C	Wright J-6-9 (R-975)	P&W Wasp C-1	Wright J-6-7 (R-760)
Horsepower	220 hp.	220 hp.	300 hp.	420 hp.	225 hp.
Passengers (Inc. Pilot)	5	5-6	6	6	4
PERFORMANCE					
High Speed	125 mph	125 mph	138 mph	150 mph	130 mph
Cruising Speed	105 mph	105 mph	112 mph	120 mph	110 mph
Climb (feet per min.)	800 fpm	750 fpm	1000 fpm	1350 fpm	900 fpm
Service Ceiling	16,000 ft.	15,500 ft.	18,000 ft.	20,000 ft.	16,000 ft.
Absolute Ceiling			25,000 ft.	27,000 ft.	22,000 ft.
Cruising Range	700 mi.	700 mi.	750 mi.	600 mi.	650 mi.
Landing Speed	49 mph	50 mph	55 mph	60 mph	50 mph
Take-off Distance	610 ft.		400 ft.	300 ft.	300 ft.
Landing Distance			350 ft.	400 ft.	200 ft.
WEIGHT/CAPACITY					
Empty Weight	1870 lbs.	2114 lbs.	2250 lbs.	2503 lbs.	2000 lbs.
Useful Load	1430 lbs.	1590 lbs.	1750 lbs.	1780 lbs.	1300 lbs.
Gross Weight	3300 lbs.	3704 lbs.	4000 lbs.	4283 lbs.	3300 lbs.
Fuel Capacity	83 gals.	80 gals.	100 gals.	100 gals.	75 gals.
Baggage Allowance			73 lbs.	100 lbs.	90 lbs.
Pay Load	800 lbs.	895 lbs.	923 lbs.	950 lbs.	600 lbs.
DIMENSIONS/AREAS					
Length	27'9"	28'4"	28'4"	29'11"	28'0"
Height	9'10"	9'10"	9'10"	9'9"	9'0"
Span	42'0"	42'4"	42'4"	42'4"	39'3"
Wing Chord	84"	84"	84"	84"	75"
Wing Area (excl. ailerons)	270 sq.ft.	275 sq.ft.	275 sq.ft.	280 sq.ft.	220 sq.ft.
Wing Loading	11.2 lb./sq.ft.		14.5 lb./sq.ft.	15.3 lb./sq.ft.	15.0 lb./sq.ft.
Power Loading	15.0 lb./hp		13.3 lb./hp	10.2 lb./hp	14.7 lbs./hp
Aileron Area	24.4 sq.ft.	24.4 sq.ft.	24.4 sq.ft.	24.4 sq.ft.	20.0 sq.ft.
Stabilizer Area	22.0 sq.ft. *(26.0 sq.ft.)	26.0 sq.ft.	26.0 sq.ft.	27.5 sq.ft.	22.0 sq.ft.
Elevator Area	15.6 sq.ft. *(16.5 sq.ft.)	16.5 sq.ft.	16.5 sq.ft.	17.5 sq.ft.	14.5 sq.ft.
Fin Area	2.5 sq.ft. *(4.0 sq.ft.)	6.2 sq.ft.	6.2 sq.ft.	6.4 sq.ft.	6.5 sq.ft.
Rudder Area	9.0 sq.ft. *(11.0 sq.ft.)	9.7 sq.ft.	9.7 sq.ft.	11.1 sq.ft.	7.3 sq.ft.
Wheel Tread	10'0"	10'0"	10'0"	10'0"	9'6"
Cabin Length			114"	114"	111"
Cabin Height			47"	47"	50"
Cabin Width			41"	41"	41"
NOTE: * Redesigned tail surfaces					
List Price (initial)	(1928) $ 9,700 later $12,200	$12,250	(1929) $13,500 later $12,250	$16,985	(1930) $10,900 later $ 9,985

INDEX

Several names run so consistently through the Ryan Brougham story that individual page references are omitted. Reference to the Contents Page will assist in locating general areas of participation by the following principals: Lindbergh, Charles A.
Mahoney, B. Franklin
Ryan, T. Claude